THE LOGIC OF BEING

THE LOGIC OF BEING

Realism, Truth, and Time

Paul M. Livingston

Northwestern University Press
Evanston, Illinois

Northwestern University Press
www.nupress.northwestern.edu

Printed in the United States of America

10 9 8 7 6 5 4 3 2 1

Library of Congress Cataloging-in-Publication Data

Names: Livingston, Paul M., 1976– author.
Title: The logic of being : realism, truth, and time / Paul M. Livingston.
Description: Evanston, Illinois : Northwestern University Press, 2017. | Includes bibliographical references and index.
Identifiers: LCCN 2017000585 | ISBN 9780810135208 (cloth : alk. paper) | ISBN 9780810135192 (pbk. : alk. paper) | ISBN 9780810135215 (e-book)
Subjects: LCSH: Ontology. | Realism. | Space and time. | Heidegger, Martin, 1889–1976.
Classification: LCC BD331 .L588 2017 | DDC 111—dc23
LC record available at https://lccn.loc.gov/2017000585

The paper used in this publication meets the minimum requirements of the American National Standard for Information Sciences—Permanence of Paper for Printed Library Materials, ANSI Z39.48-1992.

To Audrey

Contents

Acknowledgments

This book is the outcome of seven years of research and writing. It has been generously supported by the Alexander von Humboldt Foundation, which granted me an initial research visit in Freiburg im Breisgau in 2007, followed by two further renewal research stays in 2009 and 2013. In addition to the foundation, I would especially like to thank my faculty sponsor in Freiburg, Prof. Dr. Günter Figal, for his gracious support and ongoing discussion of the project over these years.

A portion of chapter 5 was previously published as part of "Realism and the Infinite" in *Speculations: A Journal of Speculative Realism* 4 (2013): 99–117, and a small portion of chapter 6 was previously published as part of "The Sense of Finitude and the Finitude of Sense" in *Semantics and Beyond: Philosophical and Linguistic Investigations* (De Gruyter, 2014), 161–84. I would like to thank the editors of these publications for permission to reuse this material here.

There are many people who have contributed centrally and essentially to shaping the book in its current form. First, I would like to thank Jon Cogburn, who, acting as a referee, gave the book's penultimate draft an extremely close reading and offered many suggestions for changes and improvements, almost all of which have been reflected in the final draft. I would also like to thank Lorenzo Chiesa and one anonymous referee for their helpful and detailed readings and reports. Many others have read part or all of the manuscript and have provided essential suggestions and comments: these include (with apologies to anyone I have unwittingly omitted) Kelly Becker, John Bova, John Bussanich, Andrew Cutrofello, David Espinet, Adrian Johnston, Ann Murphy, Achim Oberst, Marco Piasentier, Idris Robinson, Timothy Schoettle, Joseph Spencer, Gallina Tasheva, Iain Thomson, and Simon Walker. I would like to thank audiences at colloquium and conference presentations in Albuquerque, Freiburg, Sydney, Irvine, Provo, Colorado Springs, Evanston, Łódź, and IloIlo, and summer schools in Pittsburgh and Grado, where material that eventually found its way into the book was initially presented, as well as the participants in a graduate seminar titled "Theories of Truth" held in Fall 2012 at the University of New Mexico. Fabio Gironi, Agon Hamza, and Gallina Tasheva invited contributions to journals and anthologies which resulted in some of the material included here, so I would like to thank them for their kind invitations and support. I would like to thank John Bova, Walter Brogan, Raoni Padui, and Iain Thomson for discussions of Heidegger, over many years, which have proved essential to the shape of the treatment here. As always (and once again) I would like to thank Elizabeth Amberg Livingston for her love, support, and many conversations crucial to the development of the ideas in this book. Finally, I would like to thank Andrew Cutrofello for his ongoing and crucial support of the project, and Jane Bunker, Maggie Grossman, Nathan MacBrien, and Christi Stanforth at Northwestern for their help in marshalling it through the publication process.

Preface

The aim of this book is to explore and develop several interrelated problems about logic, sense, truth, and time. These problems are "ontological" in that they are implicated in what Martin Heidegger specified as the questions of the sense, and the truth, of being. For this reason, I have engaged closely here with Heidegger's thought and work. Nevertheless, this book is not primarily an exegetical book on Heidegger. Rather, it is an investigation into problems that, as I will argue, already posed themselves to Plato and Aristotle and arise again, in a different light, not only in Heidegger's texts but also in other philosophical discourses and projects of the twentieth century, including those of "analytic" philosophy. In developing these problems as they figure in contemporary philosophy, I also do not attempt to solve or resolve them in any uniquely final way. The main aim is just to elucidate and clarify them as they give themselves to be thought today.

The most central problem taken up here is that of the relationship of *thought* to *time*, whereby both being and becoming are to be thought and understood within the *life* of a being that is itself temporally situated. As I argue, this problem already motivated Plato's research, in the *Sophist*, into the specific sense of being and nonbeing. It led him, therein, to formulate the first logical theory of the articulate predicative unity of a sentence, thereby allowing change and nonbeing, against Parmenides's strictures, truly to be thought and said. Not only Heidegger's ontological questioning, but also the twentieth-century analytic tradition's investigations into the nature of linguistic meaning and the problematic results of formal and metalogical investigation into the structure and limits of rules, processes, and procedures, illuminate, renew, and deepen this problem, as I shall argue, under contemporary conditions. In the investigations to follow, I accordingly consider the implications of these approaches for the clarification of the *sense of being* as it is related to thought, and of the *phenomenon of truth* in its specific relationship to *time*.

Some main results of the investigation are as follows. First, the question of being, if it is to be taken up anew today, must be posed in relation to the specific structures of the *infinite and transfinite* as these are also understood in contemporary logical and mathematical thought.[1] Second and relatedly, this posing itself motivates an *ontological* realism—that is, a realism about *being*—which can be further supported and developed through a reflective consideration of the formalization of the logic of truth. Third, this ontological realism is, in turn, the necessary condition for a *temporal* realism that illuminates and clarifies the structure of the time of the world, overcoming any conception of it as determined by the constitutive capacities or activities of a thinking or representing subject. Fourth, with this jointly ontological and temporal realism, a structural basis of sense and time can be seen as indicated in the essentially *reflexive* structure of ontic-ontological difference. This structure is that of the relationship of Dasein *as* a being marked by a constitutive concern *with* being, grounded in a predetermining understanding of it, to the possible explicit retrieval of this understanding in thought.

As I shall argue, this reflexive ontological structure of *sense* and *given time* not only liberates both from the limitative strictures of a presumed finitude, but also positively indicates the specific structure of the *infinite* that subsists in both as the thinkable point of their contact with the real. This is not the real of an infinite-absolute in the singular, as it has been thought, for instance, within theology or philosophically by means of a conception of the infinite that is itself theologically determined. Rather, it is the irreducibly *plural* structure of the *mathematical* infinite, which Cantor's groundbreaking innovations delivered to rigorous mathematical thought. As I argue, that the givenness of time is determinable as infinite—in this sense—does not mean that it is given as the *eternal permanence* of objects, beings, or phenomena, or of any one or subset of them. Instead, in accordance with the constitutive paradoxes and aporias of infinite totality, it is structurally given as what can *never* be gathered into the unitary form of a total and noncontradictory presence.

The methodological guideline followed throughout the book is that of reflection on the *consequences of formalism.* Formalism and the possibility of formalization are employed both as objects of reflection and as means of indicative demonstration with respect to the structures of truth, being, and time. This does not mean, however, that I attempt to analyze claims about these phenomena in a particular formal language, or that the correctness of any specific formal calculus is either argued for or assumed. Rather, what provides the decisive methodological guideline are those results of logic and metalogic, actually paradoxical or nearly so, in which formalism itself encounters and demonstrates the *inherent limits* of its own scope of application and sense. In taking these results as demonstrative in this way, I have made an extended application of a formal methodology that Heidegger himself describes and employs, namely that of *formal indication.* Formalization, in this sense, is to be sharply distinguished from generalization or the demonstration of the maximally abstract or most universal. Instead it is an indicative methodology by means of which the *specific* and singular phenomena are shown, in each case, in just the ways in which they present themselves. In this respect, I believe the methodology developed and maintained here remains faithful to the idea and practice of phenomenology as Heidegger understands it, outside the ambit of any essential assumption or invocation of a constituting subjectivity.

In *The Politics of Logic* (2012), employing a similar guideline of reflection on the consequences of formalism, I distinguished among four formally possible orientations of thought, each determined by the specific way it conceives of the constitution and limits of the totality of thinkable beings as a whole.[2] Because of the way each orientation constitutively involves a specific conception of the possibility or impossibility of thinking beings as a whole, the problematic developed there was already essentially related to that of what Heidegger calls the ontological difference, although I did not discuss Heidegger's philosophy there. The critique of the onto-theological orientation of thought on the basis of the two post-Cantorian orientations, in particular, essentially parallels Heidegger's critique of metaphysics as constituted onto-theologically by the forgetting or erasure of this difference. But as I argued there, the metalogical results underlying the critique of onto-theology also equally provide grounds for criticizing the *constructivist* orientation that, by contrast to onto-theology, understands the totality of thinkable beings as constituted and limited by the finite powers of specifically human thought. In this book, I further develop the implications of formalism in overcoming both of the "pre-Cantorian" orientations of onto-theology and constructivism. The result is to suggest terms for overcoming not only the theological dispensation that assigns beings as a whole to a transcendent infinite-absolute, but also the anthropologism that seeks to constrain them within a constitutive analytic of human finitude.

It has been said, insightfully,[3] that one can see clearly at least one aspect of the methodological difference between "analytic" and "continental" philosophy by considering the contrast in the titles of two books written by leading practitioners of each: Michael Dummett's *The Logical Basis of Metaphysics* and Heidegger's *Metaphysical Foundations of Logic.* It is true that the difference in the relative prioritization of logical and metaphysical thought shown here points to an essential difference between distinct and methodologically characteristic senses and ways of "grounding," basing, or founding. But as I try to show here, the distinction nevertheless leaves open the possibility of a *twofold* but unitary formal-hermeneutical investigation into the factically given phenomena themselves. In this investigation, on the one hand, specifically *logical* structures and results are interpreted with a view to their "metaphysical" (or actually ontological) implications, while on the other, the foundations of "logic" (in a broad sense) are interrogated in relation to the formally indicated questions of the sense and truth of being. This method, I suggest, captures *one* way (there are many others) in which some of the best insights of both traditions can be accommodated within a single theoretical framework.

The investigation is divided into two parts, the first focusing on the structure of truth and the second on that of time. The first part develops a unitary interpretation of the truth of sentences as doubly founded, both *ontologically* in the deeper underlying sense of the being of entities, and *formally-logically* in the logical structure of a language.[4] This provides an ontologically based critique of the long-standing assumption, decisive in Frege's picture of language and in the early motivation of analytic philosophy, of the essential *timelessness* of logically articulated contents of language and thought. It thereby motivates a formally based account of the temporal character of natural languages, as they change over historical time, as arising from a formal and fundamental structure of *undecidability* at the basis of instituted sense. In part 2, the broader ontological and temporal implications of this formal structure are developed according to guideline of an *ontological realism.* This realism is predicated on the maintenance of *bivalence*—the requirement that each statement be either determinately true or false (or both)—with respect to ontological and temporal claims. This requirement leads, I argue, to a realist and paradoxical account of the ontological basis of time as it is "given" to be experienced and thought. Finally, I argue that the elucidation of this temporal structure can provide a critical perspective on the constitutive problems and essential limits of the global regime of calculability and technological manipulation of entities, information, and capital which broadly characterizes our collective present.

A few words, finally, about the position from which I speak in this book. As already noted, my primary aim here is not to interpret Heidegger or his philosophy but rather to develop certain ideas and problems that are suggested therein, in dialogue with approaches and results of the analytic, and other, traditions. Many of the questions and topics discussed here were considered or formulated by Heidegger, but they are also relevantly discussed (as we shall see over the course of the investigation) by Plato, Aristotle, Kant, Nietzsche, Frege, Wittgenstein, Dummett, and Davidson, among many others. The argument I make here is also critical of Heidegger, and at deep variance with his own conclusions, in many crucial respects. In the register of politics, in particular, the conclusions I draw here, in part, from certain of Heidegger's ideas about ontology and history are very different from, and in many cases opposed to, many of the conclusions he himself appears to have drawn from them.[5] I do not speak here, thus, either for Heidegger or as a "Heideggerian," and *a fortiori*, it is no part of my aim to defend either the overall trenchancy or the legitimacy of Heidegger's thought as a whole.[6] Indeed, with respect to the narrower question of the contemporary inheritance of Heidegger's thought, my hope is only that the results of this investigation will themselves contribute to the ongoing critical discussion of the many unresolved problems his writings leave for us today.

Part 1

Truth

1

The Logic of Being: Plato, Heidegger, Frege

1. Plato: A Problem of Time and Being

In a passage in his late dialogue *Sophist,* Plato articulates clearly the interlinked problematic of logic, truth, and time in which, as I shall argue, a continuation of Heidegger's questioning of being and the legacy of the twentieth-century "analytic" philosophy of language today converge. The passage comes in the course of the Eleatic Visitor's account of the views of his predecessors, amounting to what he calls a "battle of gods and giants" over the nature of being.[1] One party (the "giants") seek to "[drag] everything down to earth," defining being in material terms "as the same as body" and holding that "only what offers tangible contact is," while the others (the "gods") "insist . . . that true being is certain nonbodily forms that can be thought about."[2] By contrast with the materialists, who despise and refuse to listen to those who claim the existence of anything noncorporeal, the gods "[take] the bodies of the other[s], and also what they call the truth, and they break them up verbally into little bits and call them a process of coming to be" or becoming, distinguishing this sharply from the being of what is thinkable in itself.[3] Whatever may be the merits of this division, even those who hold the materialist position, the Visitor argues, must at least accept the existence of mortal animals. In doing so, they accept the existence of ensouled bodies, and thus count the soul or *psuche* among what is.[4] As these "rough men" must also admit, though, souls differ in being just or unjust, intelligent or unintelligent. They must thus also admit that souls *become* just or intelligent by the "possession and presence [*hexei kai parousia*]" of justice or intelligence.

Accordingly, even if the materialist continues to maintain that all that exists is to be understood in terms of bodies or the ability to affect them, he must be prepared to give a further account of this ability, capacity, or power itself. This account will point, ultimately, to what even those who recognize nothing other than change and becoming in being must nevertheless identify as the most basic underlying characteristic of whatever is:

> Visitor: Then let's go back to questioning [the materialists]. It's enough if they admit that even a small part of *that which is* doesn't have body.

> They need to say something about what's common to [*sumphues gegonos*] both it and the things that do have body, which they focus on when they say that they both *are*. Maybe that will raise some confusion for them. If it does, then think about whether they'd be willing to accept our suggestion that *that which is* [*to on*] is something like the following.
>
> Theaetetus: Like what? Tell me and maybe we'll know.
>
> Visitor: I'm saying that a thing really is if it has any capacity [*dunamis*] at all, either by nature to do something to something else or to have even the smallest thing done to it by even the most trivial thing, even if it only happens once. I'll take it as a definition that *those which are* [*ta onta*] amount to nothing other than *capacity* [*dunamis*].[5]

The suggestion, which looks ahead in one way to Aristotle and in another to Nietzsche, identifies capacity, potentiality, possibility, or power as the underlying characteristic of all that actually is. Such a position is, the Visitor suggests, obligatory for the materialists because of their own claim that all that exists is constantly becoming; in particular, once a partisan of the position admits the existence of living souls and *their* possibility of temporal becoming, it is necessary to admit the general existence of possibility or *dunamis* itself.

The discussion now turns to the interpretation of the contrasting position of the "friends of the forms," those who sharply distinguish generation (or coming-to-be) from being, holding that we "deal with [*koinonein*]" the former through the body and senses, but with the latter only through the *psuche* and logical reasoning [*logismou*] (248a). Both kinds of access, however, whether through the body or through the soul and *logos,* obviously involve a "dealing with" or having in common [*koinonein*] which must be clarified. As for the materialists, the solution of this difficulty for the "friend of forms" turns on the *actual life of the soul,* or *psuche,* capable of touching in thought and knowledge on the real of being in itself. In particular, the Visitor suggests, one possible position for the friend of the forms is the one just sketched from the position of the materialists: that the possibility of any such trafficking or dealing is based in the power or capacity of two things to come together and affect or be affected by one another (248c). On this position, the commerce of the soul with both generation and being is again rooted in *dunamis*: power, capacity, or possibility.

As the Visitor notes, however, the friends of the forms will not initially agree with this definition. For separating being from generation as they do, they will hold that *only* generation (or coming-to-be) involves such a power of body and matter to affect or be affected; being, on the other hand, does not "fit [*harmottein*]" with it (248c–d). It is here, though, that the Visitor invokes a consideration that will prove decisive in establishing the actual possibility of a *coexistence* of change and motion with being in itself: that of the actual basis of the understanding of being in the temporal *life* of the being that thinks (248e–249b). According to the conclusion the Visitor quickly draws, to deny that "change, life, soul, and intelligence" are actually present in absolute being would be to "[admit] something frightening" and untenable; hence, it must be acknowledged that both change and that which changes genuinely take part in being (248e–249b).

Whether or not one can recognize in the position of the "friend of forms" the actual position of Plato himself, or perhaps some earlier version of it, it is clear that what is at issue in the Visitor's challenge is the deep temporal problem of what Plato elsewhere calls *methexis* or participation. This is the problem figured elsewhere in Plato's corpus as that of the relationship between the always-changing objects of the senses and their timeless and unchanging but thinkable forms. Its setting here in the context of the battle of the materialists and the formalists, however, effectively poses a new and sharpened challenge to the defender of the existence of static forms. For through the Visitor's argument, the friend of forms is forced to admit that there must be some real *relationship* between the temporal realm of becoming and the static realm of thinkable beings in themselves. For this reason, it is no longer possible to relegate what is changing and becoming to simple nonexistence: change and becoming must themselves *also* be admitted as part of what ultimately is. The consideration that most directly demands this admission is that the living, dynamic soul nevertheless has the capacity to *know* or understand being or what is. The problem of this capacity is thus the problem of the *temporal structure* of the *thought of being* as such, or of the means by which a being irreducibly situated in time nevertheless grasps the categories and underlying structures of whatever truly is.[6]

Posed this way, the problem puts significant pressure on the formalist's claim to distinguish the realm of what is changing from that of the timeless and unchanging forms. But if the Visitor's argument is correct, it arises just as deeply for the materialist, who does not admit the existence of anything timeless or eternal, holding that *everything* is always changing and becoming, as for the adherent to (some version of) the "Platonist" doctrine of timeless forms. For on *both* of the opposed views, according to the Visitor, in order to account for the categories in which *we* understand entities and phenomena, it is necessary to admit the privileged existence of a living *psuche* capable of thinking or knowing being as it is in itself. Both parties, moreover, conceive this *psuche* as the nexus of a capability of *combination* or *mixing in copresence* which ultimately demands, on either view, a real relationship of coexistence and interaction between the changeable and moving, on one hand, and being as it is in itself, on the other. But what is the nature and structure of such a *dunamis*, or capacity, which lies (as both parties understand it) at the basis of both the definition of what it is for anything *to be*, and the possibility of the being of things to be *understood* and *known*?

Seen in most general terms, the problem, visibly determined by the joint inheritance of Parmenides's arguments against temporal change and becoming and Heraclitus's insistence on them, is essentially the question of the relationship of time and change to *truth*. This is the question of the formal or ontological basis on which a mortal being, itself situated in time, can nevertheless touch, in thought, on the reality of what truly *is*. Its broadest horizon is the question of how truth itself "is" temporally: what kind of temporality characterizes it as such, and how this temporality is related to the time in which it is thought and understood by finite and mortal actors. Understood in this way, the problem can also be understood as the problem of *realism* about the intelligible *sense* or *meaning* of objects, phenomena, and entities themselves. As the Visitor argues, both the materialist and the formalist must, to ground their respective positions, recognize a basis of the possibilities of things and entities in the structure of *what is*. But the possible knowledge of this basis amounts, for both positions, to the possible knowledge of the *basis* of their appearance, manifestation, intelligibility, presentation, or representation in the structure of what is most deeply real in itself. If the materialist as much as the formalist, as the Visitor argues, must acknowledge such a possible knowing of being, then there is already implied in both of their positions some relationship—obscure though it might remain—between the temporally situated life of the embodied soul and this real "in itself" at the basis of the meaning and intelligibility of things. The question of the status of this relationship is subsequently developed in modern philosophy, from Descartes to Kant, as the primarily *epistemological* problem of the *finite subject*'s possible knowledge of the world. As we shall see in subsequent chapters, though, the Kantian critical setting of the problem within that of the relationship of the sensible to the intelligible provides the conditions for it to be transformed and for its deeper *ontological* (rather than simply epistemological) significance to be recovered. Thereby, it becomes (once more) a problem of the relationship of *sense* to *time*: how are both the sensory and the conceptual givenness of things related to the specific structure by which time is itself given to the experience and thought of a temporally situated being?

From the perspective of the Visitor (if not necessarily that of Plato himself), the suggestion that being and becoming combine in the privileged medium of the soul's capacity represents just the beginning of a general account of the relationship obscurely indicated by Parmenides himself in the remark that "thinking and being are the same."[7] Nevertheless, in the context of the dialogue, this suggestion will provide the basis for the Visitor's resolution of the formidable Parmenidean problem of the possibility of saying "what is not." This is the problem that if to speak at all is to say *something,* then it is apparently impossible to say what is not, since *what is not* is apparently not something.[8] But to say of something that it changes is apparently to say that it is *not* now something that it has been, or that it was earlier *not* some way that it is now. Any discussion of change and becoming thus appears to involve what Parmenides argues is impossible: that something can coherently be now said or asserted about what is not, either what is no longer or what is not yet. But the form of solution that the Visitor suggests provides at once the sketch of an answer to the question of this possibility and a rebuttal of Parmenides's argument against the possibility of saying what is not itself. In particular, once it has been admitted by all parties that some knowledge of *being, rest,* and *change* is somehow possible, it must also be possible (so the Visitor argues) for these types to enter into differential logical or conceptual relationships with one another in the medium of the soul. One must thus acknowledge also, in addition to the initial three great "types," the *sameness* and *difference* of these.[9] It is then possible, the Visitor argues, to consider that *difference* can mix with all of the other great types, and in particular that a mixing of *difference* with *being* produces "something different from" it: that is, something that is *not* being, in the sense of something *different from* what is.[10] This allows, finally, the actual essence or form of "that which is not" (the *me on*) to appear, and allows it finally to be said that it blends with speech or the *logos* in the deceptive discourse of the sophist.[11]

The actual form and structure of the potential combination of types in the soul which solves the problem of the relation of being and becoming is left obscure, for now. But as the dialogue moves toward its conclusion, the Visitor suggests a superior *logical-syntactical grammar* of essential types figured (in a more than simply metaphorical sense) by the actual relations between phonemic or lexical elements characteristic of language. The suggestion is closely related to the methodology of "collection and division," or *synthesis* and *diaeresis,* suggested by Plato in a number of late dialogues. Here, it depends also on the recognition, common to the materialist and the friend of forms, of an essential capacity for synthesis or combination rooted in the nature of the soul. In particular, given such a possibility of combination in the soul, the Visitor can argue that such opposite types as *change* and *rest,* even if they cannot associate with one another, must themselves have some capacity for association [*dunata epikoinonein*] with other general types, including particularly *being* itself (251d–252d). All of those who discuss being and becoming, the Visitor argues, are forced in their very statements to "use *being* about everything, and also *separate, from others, of itself,* and a million other things" (252c). Even those who hold that everything is in motion, in using the signifier "is," thus admit some possibility of the mixing of being and becoming. In this way, they incessantly "link . . . together . . . in speech" the various types and attributes determinative of objects and phenomena, and so concretely exhibit the actual mixing and combination of these types in their discourse.

Since it is untenable to suppose either that all of the types or elements mix with one another or that none do, it is necessary to recognize the existence of a broader structural determination of the actual possibilities of mixing, and of a specific type of art, or expertise, capable of discerning these possibilities and relationships. This expertise is the art of grammar.[12] As is the case, similarly, with the capacity to combine high and low notes to produce pleasing harmonies, the *techne* (or expertise) of grammar requires a specific kind of knowledge of the capacity of some kinds to blend and others to refuse such combination. This is a knowledge, according to the Visitor, proceeding through *logos* and allowing the possibility of recognizing those types or elements that run through everything else. It involves a singular knowledge of the number and relationships of the forms; such knowledge, the Visitor suggests, may in fact be that characteristic of the philosopher, the knowledge that "free people have."[13] According to the Visitor, it is in fact none other than the science [*episteme*] of dialectic.[14]

The dialectical method, as the Visitor then develops it, is thus closely related to the possibility of a *counting* of the most general types of being, as well as of being and nonbeing themselves in relation to what they characterize. In particular, a brief elucidation of the structure of mixing and separation among the "most important" types leads the Visitor to identify the five great types: *being, rest, motion, identity,* and *difference.* Difference "permeates them all" in allowing each to distinguish itself from the others. Motion, in particular, is "other than being" in that it is different from being; with respect to motion, thus, "nonbeing is," although as has been shown, motion, as something that *is,* also *takes part* or *participates in* being.[15] In fact, this consideration holds generally; in "each of the types," difference is operative to make them "other than being, and therefore nonbeing." This allows the Visitor to specify the duality of being and nonbeing, with respect to what each characterizes, as the problematic relation between a *multiplicity* and an *infinitude*: "And so, in relation to each of the [ideas], being is many, and not-being is infinite in number."[16]

Thus the realm of the possible application of the dialectician's art, that of counting the types of being by discerning the ideas, is here set off against the indefiniteness, unlimitedness, or *apeiron* character of nonbeing. It is in this infinitude of nonbeing that the synthetic/diaeretic technique of the dialectic reaches the specific limit of its capacity to discern forms.

The relationship that the Visitor asserts between the forms, as available to the philosopher possessing the grammatical *techne* of dialectic, and the underlying structure at the root of the coherence of ordinary discourse, can be seen as inaugurating the project of a *logical analysis of ordinary language.* The Visitor sees such an analysis as capable of discerning, in the characteristic structures of ordinary speech, the implicit or presupposed structure of the ultimate types defining the thought and knowledge of being in itself. Though chronologically preceding what is usually recognized as the first development of "formal logic" in Aristotle's syllogisms, this suggestion can be seen as defining a more original Platonic conception of (what is already here understood as) *logical form.* In insisting on a real basis of the linguistic *logos* in the life and temporality of the embodied soul, the Visitor suggests a basis of the possible knowledge of being in the dynamic logical capacity of the soul to gather diverse types into the unity of a rational discourse.

This conception leads the Visitor also to articulate what has been seen as the first significant conception of the grammatical structure of a *predicative* sentence in the entirety of the Western tradition. Specifically, (261e–262c) a sentence is a combination of names and verbs; it is no more possible for a series of mere names to come together to form a significant sentence than it is for a series of verbs to do so. Rather, someone who utters a simple sentence such as "Man learns" puts together a noun with a verb; in so doing, he "gives an indication [*deloi*] about [*peri*] what is, or comes to be, or has come to be, or is going to be" (262d). Such an indication does not simply consist in naming either things or actions. Instead, the "weaving together" of names and verbs into a sentence constitutes what is itself an accomplishment, that of *saying something.* The discernment of the specific combinatorial possibilities of these elements thus makes possible the very structure of the *logos* itself, the possibility of saying or asserting rather than simply naming. And it is this structure, as well, which establishes that every sentence must be *about something* [*tinos*]; formed by the combination of a significant subject or name for an object and the sign for an action, the sentence as such thus has the articulate structure of *saying something about something.* It is through this capacity that it is finally possible for a *logos* to be false or to say "what is not"; in particular, a false logos says about its subject something *other* than what is (263a–b).[17]

The Visitor's rigorously developed solution to the specific problem of the sophist thus points to two more general problems of logic and sense, characteristic of the development of reflection on linguistic meaning up to the present. The first is the problem of the nature of *predication.* This is the question of the specific structure of sentences, propositions, assertions, or judgments, such that something is said, asserted, or judged of or about something else. The second is the problem of linguistic *intentionality*: that of the link between speech and the external reality that it characterizes, describes, denotes, or refers to.

Whether or not the Visitor's solution can be considered adequate in addressing these problems, it is significant for their subsequent development that linguistic sense is here thought as arising from the superior structure of an *a priori* connection of types. The temporal mode of this connection is that of the life of the rational *psuche,* or the capacities of the *living being* capable of specifically *logical speech and thought.* The conception looks forward to a problematic of the relationship of logical forms to the commonality of a shared life that is as broadly representative of twentieth-century philosophical inquiry into language as it was already for Plato. Within this inquiry in both its "analytic" and phenomenological/ontological forms, the question that arises repeatedly and decisively determines methods, considerations, and results is that of the particular mode of the relationship of the *logical form of language* to the *facticity of a life.* As such, it is the question in the ultimate horizon of which it is possible to envision a joint contemporary inheritance of the legacy of an "analytic" inquiry into the specific rational structure of language, and that of an *ontological* inquiry into the structure and meaning of being.

2. Heidegger: The *Sophist* and the Onto-theo-logy of Being

Martin Heidegger's discussion of the "battle of gods and giants" over being, in his comprehensive Marburg lecture course of 1924–25 devoted to the interpretation of Plato's *Sophist,* comes just after he briefly discusses the remark of the Visitor which he would use, roughly two years later, as the epigraph for *Being and Time*:

> For manifestly [*delon*] you have long been aware of what you mean when you use the expression "*being*." We, however, who used to think we understood it, have now become perplexed.[18]

In its original context, the remark is addressed to Parmenides and all those who have earlier attempted a "critical" definition [*epi krisin . . . diorisasthai*] of the number and types of beings [*ta onta*].[19] These predecessors, the Visitor avers, have been "careless," and have told stories [*muthon*] as if to children, rather than considering whether their arguments [*legousin*] can genuinely be followed out. In particular, Heidegger suggests, in aiming to deal with being, they have instead told stories about beings or entities, actually only explaining beings in terms of beings; thus they have "moved naively in the dimension of beings" and thereby failed to enter into the "dimension of the *Being* of beings."[20]

Heidegger here alludes to the guiding distinction that, though it would be explicitly formulated only two and a half years later,[21] already in 1924–25 provides an essential guideline for his philosophical interpretation of Plato as well as the substantive and still unresolved problem that he sees as posed in Plato's text. This is the *ontological difference* between beings, entities, or *whatever is*, on the one hand, and their *being*—the fact or circumstance *that they* are, or the precondition of this—on the other. For Heidegger already in 1924–25, a more or less explicit understanding of the ontological difference is essential in order for a question concerning the meaning and significance of being to be raised. Accordingly, it is equally essential that *being* not be understood simply in "ontic" terms: that is, in terms of one or another entity. For the explanation of being in terms of beings, since it presupposes what is to be explained, does not succeed even in raising the question of the significance or meaning of being itself (let alone answering it).

Here Heidegger marks Plato's awareness of the distinction (at least in a partial and partially obscured way) in the specific terms of the Visitor's criticism of his predecessors. The implications of the ontological difference in the Visitor's discussion are also intimately related to those of the possibility of *saying* something or *speaking* about being itself. In particular, according to Plato as Heidegger reads him, all of those who have tried to explain being by counting the types of the most proper beings, including monists like Parmenides himself as well as those who account for beings in terms of a duality or plurality of principles, have necessarily invoked in these accounts the specific structure of *legein*, the activity or result of *saying* or *speaking*. In so doing they have already, according to Plato as Heidegger understands him, also invoked "something else" that is "co-said" in *all legein*, or in all speaking about entities, insofar as the *logos* itself includes the possibility of its *own* saying as a constitutive and implicit moment. And this "something else" that is inexplicitly "co-said" along with any description or discussion of entities, Heidegger says, is "no less than Being itself."[22]

It is thus that the Visitor comes to pose the question of what is really meant by the sign or word "being," which all of those who discourse about beings and their underlying structure ceaselessly presuppose. In particular, Parmenides and the Eleatics have proposed the hypothesis "*hen to pan*," or "All (that is) is one." Plato's demonstration, Heidegger suggests, does not have the goal of simply disputing this hypothesis but of showing that in it "there resides a moment which reaches beyond its own proper sense."[23] Specifically, the hypothesis raises the problem of the signification of the *term* or name "being" in relation to what it signifies. Here, invoking a problem that is more fully developed in the dialogue *Parmenides* itself, the Visitor states an inherent dilemma involved in assuming "being" to have such a signification within the scope of the Parmenidean hypothesis. The dilemma is that the admission of a *name* or term for being, *in addition to* what is named or signified itself, already requires that *there are at least two,* and not only one, as the hypothesis apparently requires. For in addition to whatever is, there is now also (at least) the name or term itself. Far from being a mere sophistical or pseudo-problem, according to Heidegger, the problem for the hypothesis of the unity of being in fact concerns the very structure of the *logos* as *legein ti,* or as a saying *of* something "about something." The fact that every *logos* is a *legein ti,* that every *logos* is about something, here means that the hypothesis, as a *logos* about being, says *of* being that it is one. Heidegger accordingly suggests that the Visitor's inquiry as to what is meant by the word "being [*on*]" captures the "genuinely central concern . . . of the whole dialogue"; this, in particular, is the concern to *prepare* the ground for a possible ontology by posing the question of the meaning of being by interrogating the specific relation of the meaning of the *term* "being" in relation to the *logos* that articulates it.[24]

The nature of what is thereby brought out emerges in the course of the Visitor's questioning directed toward the partisans of materialism and those of the forms. The materialists, in particular, understand "body" and "material thing" to signify the same as *ousia* (or "being" in the sense of the [nominalized] present participle of the verb *einai,* "to be") (246a–b), holding that everything is constantly in motion and becoming. The friends of forms, by contrast, identify *ousia* with *eidos,* privileging the static form or idea as that which is most genuinely or substantially real. In having to admit the existence of *phronesis* (knowledge or understanding) in the actual living soul, the materialists necessarily admit a more general basis in the structure of the soul for the possibility of combination of the visible with the intelligible. This more general basis is the *dunamis,* which, the Visitor suggests to the materialist, might thus be taken as the basic structure of everything that is. For the formalists, by contrast, true being, which is understood as *stasis*, is known only through thought and reasoning. Nevertheless, the reality of knowledge as a capacity of the soul again forces the adherent of the position to acknowledge a more general possibility of combination or being-with-one-another.

Heidegger suggests that the common form of this solution, for both parties, directly captures Plato's own solution to the difficult problem of the communion of change and becoming with *stasis*:

> Being itself, then, will mean for Plato, if he is to make both these positions intelligible, *dunamis*, as *the possibility of co-presence with something* [*Möglichkeit zur Mit-Anwesenheit bei etwas*], in short *dunamis koinonias*, or in a fuller determination, *parousia dunameōs koinonias*, factual occurrence of the possibility of being with one another [*Vorhandensein der Möglichkeit zum Miteinandersein*] . . . This concept of *dunamis koinonias*, as the possibility of being with one another, is the focus of Plato's entire ensuing discussion.[25]

In proceeding from the two positions discussed to the discernment of a characteristic unified phenomenon, Plato (according to Heidegger) identifies the possibility of knowledge as a "particular *koinonia*," or being-in-common. This *koinonia* "includes, in the first place, a connection, a companionship [*ein Zussamenhang, ein Mit-sein*] of the *psuche*, of *nous*, with the *eide*, i.e. a being of *genesis* with that of the *aei on*."[26] According to this solution, specifically, "the soul is the being in which we can see that in fact *stasis* is co-present with movement."[27]

But what is shown in the way Plato places exactly this requirement on the form of an account of the proper mode of access to genuine beings, according to Heidegger, is the inexplicit but guiding presupposition of a specific understanding of the meaning of being [*ousia*]. This presupposition guides not only Plato's questioning but the whole ontological problematic of the Greeks:

> This meaning of Being does not naturally lie in the light of the day but instead can be understood explicitly only by means of a subsequent interpretation [*Auslegung*]. The meaning of Being implicitly guiding [Greek] ontology is Being = presence [*Sein* = *Anwesenheit*]. The Greeks did not get this meaning of Being from just anywhere, they did not just invent it, but rather it is the one borne by life itself, by factical Dasein, insofar as all human Dasein is interpretative [*Auslegendes ist*], interprets itself as well as everything that is a being [*alles Seiende*] in whatever sense. In this interpretation there is operative [*lebendig ist*] an implicit sense of Being. And indeed the meaning of Being, as the Greeks inex-

> plicitly understand it, is drawn out of the natural immediate interpretation of Being by factical Dasein: Being means: to be there already at the very outset [*im vorhinein schon da sein*] as possession, household, property [*Anwesen*]—put more sharply: as presence [*Anwesenheit*]. We will make use of this meaning of Being (which we ourselves first make visible, although of course we cannot discuss it further in this context), namely Being = presence, because *in it lies concealed the whole problem of time and consequently the problem of the ontology of Dasein.*[28]

The problem posed in the battle of gods and giants thus captures in a very specific way the problem of the ultimate basis of the being of entities as it connects with the problem of time. In particular, according to Heidegger, the problem of the meaning of being in general here appears specifically as the problem of the most proper mode of presentation of *what is* in itself, a problem that is already posed with the admission, made by both parties to the battle, that the soul's knowledge of being is in some way possible. The solution points, according to Heidegger, to "what today we would call a *consideration of essence* [*Wesensbetrachtung*] or a *knowledge of the apriori*."[29] That is, in locating the specific possibility of a combination of *stasis* and *kinesis* in the structure of the soul, Plato develops what is actually an original account of the possibility of *a priori* knowledge, which is now treated, Heidegger notes, in the phenomenological theory of eidetic cognition.[30]

Although Heidegger urges that the Visitor's solution should not be understood as a subjectivist one, it nevertheless depends, according to Heidegger, on Plato's problematic assumption that "the grasping of the *apriori* resides on the same level as the grasping of the ontical in general."[31] In particular, the factical and substantial existence of the *psuche*, as the basis for the possible combination of the essential types of *stasis* and *kinesis*, is invoked here as the basis for the possibility of any knowledge of being as such. At the same time, though, the priority of the *a priori*, and the whole possibility that the soul's synthetic capacities can indeed yield knowledge of being, depend further on the specific relation of the soul to the structure of *logos*. Plato understands this relation as one of *possession* or *having*: the soul can know and understand the "logical" structure and character of beings because it itself *has* or *possesses* the structure of the *logos*, in some (more or less obscure) sense. This having or possessing is also thought by the Visitor as a distinctive logical *capacity*: the soul's rational *dunamis*, which allows for the underlying logical structure of being to be grasped and understood. This having or possession subsequently receives an influential formulation in Aristotle's understanding of the human being as the *zoon logon echon*: the specific animal defined by its having or possessing the *logos.*[32] Later, it is understood as the soul's possession of reason, intellect, rationality, language, or the capacity for logical thought. In this sense, the Visitor's particular solution to Parmenides's problem can be seen as substantially inaugurating the whole subsequent philosophical development of the theme of a rational capacity for knowing the real. But in the Visitor's argument itself, this definitive possession of the logical *dunamis* which makes it possible for us to have some relationship to logic and reason is itself grounded, as Heidegger emphasizes, in a specific conception of the *temporality* and *life* of the being possessing it. This is what Plato understands as the being of the *psuche* and its capacity for logical synthesis, which allows for the limited combination of the overarching *gene* or types.

Plato thus ultimately gives, on Heidegger's reading, what can be called (in an anachronistic but accurate sense) a *psycho-logistic* theory of being and truth. In particular, it is, for Plato as Heidegger reads him, the temporal capacity of the soul dynamically to synthesize logical elements in actually speaking and thinking that ultimately accounts for the sense and meaning of being in itself, insofar as it is thinkable or knowable. The temporal contact of the *psuche with* the larger structural possibilities established by the preeminently existing types is itself guaranteed by this dynamic structure of the *psuche* as the capacity for combining them into a thinkable *logos*. The underlying structural basis for this connection between the *psuche*'s capacity to produce intelligibility and the logical structure of being in itself is the intercommunicating series of *koinonia,* or commons, that Plato sees in the capacities of the *psuche,* the structure of the predicative sentence, and the intentional relation between sentences and their objects. All of these are linked together, for Plato, in the notion of a *logical/ontological copresence* that preserves meaning and ensures the communication among the different orders of the psychic, the material, and the ideal. And the ultimate basis of this copresence is the manifold *koinon* (or common) of the *logos* itself.

As influential and determinative as this conception is for the subsequent history of philosophical thought, it is nevertheless deeply problematic, according to Heidegger. This is because of the way it already obscures the ontological difference between being and beings, and hence at least begins to make the specific *ontological* inquiry into the sense or meaning of being itself obscure or impossible. In particular, Plato's conception of the logical *koinonia* of the soul already begins to substitute for this ontological inquiry the inquiry into the specific structure of the *psuche.* With this substitution, the *psuche* becomes privileged in a twofold sense: both as the specific kind of entity whose capacities account for our possibility of knowing being as it is in itself, and as the nexus, through these capacities, of the logical articulation of things and events. When, however, both being and its articulation in *logos* are thus understood presumptively in terms of presence to the *psuche*, what is obscured is the prior basis for the phenomenon of *presencing* itself in the ontological difference between being and entities. For with the substitution, the sense and meaning of entities is no longer understood as based in this difference, but rather in terms of the capacities of what is in fact only one type of entity among others. Here, then, the question that is properly posed only on the basis of an inquiry into the whole of beings as such is presumptively answered by reference to a particular one (namely the soul) and to its still obscure capacity for logical synthesis or unification in simultaneous presence. What is thereby obscured, with decisive consequences for the subsequent development of thinking about logic, sense, and time, is the very possibility of a more basic inquiry into the basis for the intelligibility of things insofar as this basis is itself ontologically and temporally structured by the difference of beings and being.

This critique of Plato's psychologism plays an essential role in generating the broader critique of the "logical tradition" of the West, along the guideline of a radicalized questioning of the sense of being, that Heidegger announces in the first pages of *Being and Time.* As Heidegger explains there, the critique calls for a critical "destruction" of the tradition of Greek ontology on the more original basis of the question of *temporality*: that of the basic structure of time as it is given to thought and experience.[33] As such, Heidegger suggests, it interrogates in particular the ancient and continuing determination of Dasein as the *zoon logon echon* or as "that living thing whose Being is essentially determined by the potentiality for discourse [*durch das Redenkönnen bestimmt ist*]."[34] For this conception as it unfolds in the subsequent development of the Western tradition according to Heidegger, the being of the *psuche* enjoys a recurrently marked priority in the description of the ultimate basis of meaning and truth. Thus for Aristotle, for example, the soul is, "in a certain way," all entities.[35] And for Aquinas the "distinctive entity" that is the soul [*anima*] is "properly suited to 'come together with' entities of any sort whatever [*die Eignung hat, mit jeglichem irgendwie Seienden "zusammenzukommen," d. h. übereinzukommen*]" in such a way as to produce the possibility of truth and ensure its status as a "transcendental," something that goes beyond any generic classification of entities to characterize any subject matter whatsoever.[36]

If, moreover, the Platonic configuration that Heidegger interrogates in the *Sophist* lectures already provides a definitive critical target for the "destruction" undertaken in *Being and Time*, the *logical* problematic thereby elicited retains a priority for Heidegger's deconstruction of metaphysics throughout its whole itinerary. This remains so even, and even in a heightened form, after Heidegger's thought undertakes the famous "turning" in the 1930s from what he specifies, in the *Beiträge zur Philosophie,* as the "guiding question" of the being of entities to the historical "grounding question [*Grundfrage*]" of being's (or "beyng's") truth.[37] In the altered conception of the *history of being* to which this leads, the being of entities is said to be successively "metaphysically" determined according to a series of privileged figures, for instance as *idea,* as *hupokeimenon* and *dunamis,* as *transcendens,* as *cogito* and representing and willing subject, and as will to power and technology.[38] Each of these determinations grounds the intelligibility of entities *as a whole* by reference to what is thought as the most essential aspect of their character as entities. Throughout all of these specific determinations, however, the specific structure of metaphysical thinking is visible in its claim to *unify* beings as a whole into a determinate configuration of intelligibility. And according to Heidegger, the specific basis of this claim, through all the various configurations of the metaphysics of the West, is the privilege of the *logos* and the particular kind of unity that characterizes it, as the ground of the thinkability of entities as such.

This privilege of the *logos* throughout the succession of differing metaphysical interpretations of the being of entities is marked in the double hyphenation of what Heidegger specifies in a late (1957) lecture as the "onto-theo-logical constitution of metaphysics." As Heidegger here explains, metaphysics in its thought of beings as a whole always has a twofold determination, as ontology and theology, uniting the ontological determination of "beings as such in the universal and primordial [*das Seiende als solches im Allgemeinen und Ersten*]" with their theological determination in terms of "the highest and ultimate [*im Höchsten und Letzten*]."[39] But the two kinds of determination of the being of beings (from "above" and from "below") are themselves further linked in the thought of the *unity of being and thinking* that makes possible the intelligibility of beings as a whole for each specific, epochal configuration.

Heidegger finds the basis of this unity, as it is thought throughout Western metaphysics, in the specific structure of the *logos* as a gathering and unification that grounds:

> Being manifests itself as thought. This means: the Being of beings reveals itself as the ground that gives itself ground and accounts for itself. The ground, the *ratio* by their essential origin are *logos,* in the sense of gathering and letting-be [*im Sinne des versammelnden Vorliegenlassens*]: the *Hen Panta.*[40]

Thus, the two interlinked structures of ontology and theology that together define Western metaphysics have a further and still unthought unity in the structure of the *logos* itself as the grounding basis for the *gathering*, binding, and presentation of beings in the unity of a whole:

> Ontology . . . and theology are "Logies" inasmuch as they provide the ground of beings as such and account for them within the whole. They account for Being as the ground of beings. They account to the *logos*, and are in an essential sense in accord with the *logos*-, that is they are the logic of the *logos*. Thus they are more precisely called onto-logic and theo-logic. More rigorously and clearly thought out, metaphysics is onto-theo-logic.[41]

Because "metaphysics responds to Being as *logos*," it is "accordingly in its basic characteristics everywhere logic" and in particular "a logic that thinks of the Being of beings."[42] In each of the determinate configurations of metaphysics up to the present, the possibility of thinking beings as a whole in determinate figures of unity is, in fact, itself made possible by the specific kind of unity that the *logos* is thought to have. In each of these configurations, "logic" thus characterizes, at once, the specific structure of unitary thought about entities *and* the ultimate basis for the determination of the properties of these entities themselves. As such, it is, according to Heidegger, "the name for that kind of thinking which everywhere provides and accounts for the ground of what is, as such and as a whole [*das Seiende als solches im Ganzen . . . ergründet und begründet*] in terms of Being as the ground (*logos*)."[43]

If, then, the unifying and grounding character of the *logos* provides an original basis for the various simultaneous determinations of thinking and the being of entities that comprise onto-theo-logy, Plato's thought of logical form as permitting the specific *capacity of combination* of *logical elements in the soul* originally determines the conception of "being as presence" that underlies it. This Platonic conception understands the temporality of the relationship between being and entities as grounded in the logically constrained *copresence* of the thinkable in the medium of the *psuche*. It underwrites this understanding by pointing to the psychologistic correspondence that the Visitor introduces. This is the correspondence between the psychological form of life of the animal possessing the capacity of *logos* and the overarching logical/ontological structure of the forms or types. The unity of this correspondence then subsequently allows being as such to be determined as the unity of an ontic totality of beings, a determination which then always characterizes metaphysics or onto-theo-logy, for Heidegger. The determination forecloses the ontological difference between being and beings, in understanding thinkable being as such (despite, or rather through, the Visitor's logical-grammatical critique of Parmenides's own argument about falsehood) as the thinkable unity of a logically determined copresence toward which the Eleatic hypothesis—*hen on to pan*—originally gestured.

3. Frege: Truth, Sense, and the Logic of Judgment

As we have seen, the problematic of thinking and being to which Plato, in the voice of the Visitor, offers the solution of the logical *koinonia*, is originally the problem of the specific *temporality* of logical thought insofar as it captures what is thinkable in being itself. In a characteristically trenchant methodological passage in the *Grundgesetze der Arithmetik*, Frege takes up the consequences of the same problematic as it bears on the issue of truth:

> For me, truth is something objective, independent of the judging subject, for psychological logicians, it is not . . . We can capture this more generally still: I acknowledge a realm of the objective, non-actual, while the psychological logicians take the non-actual to be subjective without further ado. Yet it is utterly incomprehensible why something that has being independently of the judging subject has to be actual [*wirklich*], i.e. has to be capable of acting [*wirken*], directly or indirectly, upon the senses. No such connection between the concepts [of objectivity and actuality] is to be found . . .
>
> Since the psychological logicians fail to appreciate the possibility of the objective non-actual, they take concepts to be ideas [*Vorstellungen*] and thereby assign them to psychology. But the true state of affairs asserts itself too forcefully for this to be accomplished easily. And hence a vacillation afflicts the use of the word "idea" ["*Vorstellung*"], so that sometimes it seems to refer to [*bedeuten*] something which belongs to the mental life of an individual and which, in accordance with the psychological laws, amalgamates with other ideas, associates with them; while at other times, to something that confronts everyone in the same way, so that no bearer of ideas [*Vorstellender*] is either mentioned or even presupposed. These two uses are incompatible . . .[44]

In the "vacillation" between the two conceptions of the meaning of the "idea," or representation, which Frege attributes to the psychological (psychologistic) logicians, one can see clearly presented once more the main components of the Eleatic Visitor's psychologistic conception of the logical/psychological *koinonia* in the *dunamis* of the soul. Here, as for the Visitor, the positive basis of content or sense rests, on the one hand, in the preexisting logical structure of categories or types, in general and independently of the mental life of any specific thinker, and also, on the other, in the specific possibility that they or their representatives "amalgamate" or "associate," within the course of such a life, in the privileged medium of the *psuche* according to psychological (rather than logical) laws.

Frege's critique here consists in pointing out the incompatibility of these two components of the picture shared by the Visitor and the contemporary psychologistic logicians. This incompatibility is ultimately *temporal* in character, since it concerns the underlying question of the temporal status of logical structure itself, and its relationship to the effectivity of spatiotemporal phenomena and causal processes. But the basis of Frege's own conception of the specific mode of objectivity of the logically articulated concept, like that of much else in Frege's thought, is the essential link between logic and truth. In particular, if logic is, as Frege says near the beginning of an 1897 work, "the science of the most general laws of truth," this is because its task lies in saying "what holds with the utmost generality for all thinking, whatever its subject matter."[45] In this way the "word 'true' can be used to indicate" the goal of logic, as "good" points to the goal of the study of ethics, or "beautiful" that of aesthetics.[46] As Frege suggests in a roughly parallel passage of the late (1918) article "Thought," if "the reference [*Bedeutung*] of the word 'true' is spelled out in the [logical] laws of truth," it is thus accordingly necessary to consider logic as articulating the laws governing *what is* true, rather than "the laws of taking things to be true or of thinking" as a process or activity.[47] It is thus necessary, in order to preserve this definition of logic as articulating the laws of truth, to recognize also the privileged link between logic in this sense and being in the sense of *what is*, as opposed to the illusory or false objects of opinion, appearance, or fallible judgment. Recognizing this link is, in turn, sufficient, as Frege says in the *Grundgesetze* passage, to break any presumed link between the being of what is and "actuality" in the sense of the sensible, or of whatever has the capacity to affect or be affected.

A basis for this way of understanding the connection of logic and truth can be found in the argument that Frege gives, both in the 1897 "Logic" and in "Thought," for the conclusion that *truth is indefinable*, and in particular that it cannot be identified with any property, feature, or relationship of objects.[48] Since both the mind and its activities are conceived in terms of objects and features in this sense, the argument, if successful, suffices to defeat psychologism about truth and the laws that articulate it. The argument is that the definability of truth would imply an infinite regress. For if the truth of any sentence consisted in some property it possessed, or some relation in which it stood, it would still always be possible to ask whether it was *true* that the sentence had that property or stood in that relation. In other words, if the truth of sentences were indeed definable in terms of any property, relation, or set of properties or relations, the usefulness of the definition to determine what is involved in the truth or falsity of a particular sentence would depend, in turn, on a determination of whether that sentence indeed *had* the requisite property (or stood in the right sort of relation to something else). And the need to make this determination in each case would lead, Frege argues, to an infinite regress that would have to be traversed in order for the purported definition first to be applicable at all.

One might wonder why the regress must be vicious. In particular, a proponent of some particular definition of truth might simply hold that it is to be reiterated at each stage, without ever leading to any actual incoherence at any stage. But as Michael Dummett points out in his analysis of the argument, it actually has a further premise which suffices to show the untenability of such a picture. The premise is another claim that Frege was probably the first to make and that has elsewhere been called the "redundancy" or "equivalence" principle.[49] The claim is that the assertion "It is true that 'A,'" or the predication of the truth of A, is (in some sense) *equivalent* to the assertion "A" (or to predicating what A predicates). Frege states this equivalence principle explicitly elsewhere,[50] and given that principle, it is clear that any definition of truth in terms of any property, feature, or relation is untenable. For given any such definition, even while asserting "A," it would still be open to us to inquire whether "A" had the requisite property, feature, or relation (thus, whether "A" is true). But that this is in fact *not* open to us is a direct consequence of the equivalence principle itself.

It is suggestive that the principle at the basis for Frege's general and powerful argument for the indefinability of truth in terms of correspondence or any other notion is, at its basis, just the same one that was subsequently used by Tarski to capture what he considered the most important formal constraint on the definition of a *truth-predicate* for a particular formal language, L. For *if* applied, in this way, as a constraint on possible definitions of such a language-specific truth-predicate, the equivalence principle becomes Tarski's well-known T-schema:

(T): *s* is true-in-L if and only if *p*

where *s* is replaced with a description or name of a sentence, and *p* is replaced by the same sentence, or a translation of it.[51]

The schema would later prove decisive in many of the formal and informal approaches of the analytic tradition to problems of meaning and truth. In application to natural rather than formal languages, it would become the basis for Davidson's programmatic conception of the structure of a theory of meaning, grounded in a Tarskian truth-theory, for any particular natural language.[52] Nevertheless, as Davidson himself argues in the posthumously published text *Truth and Predication*, the applicability and definability of Tarski-style truth-definitions for any number of particular languages still leaves open a larger question, not solved by any number of such definitions. This is the question of the *general* concept of truth, or what is held in common by every language and formulated in *each* of the language-specific truth-predicates.[53]

It is in this sense, the sense of the "general concept" as opposed to particular truth-predicates for specific languages L, that truth is plausibly "normative" in that it provides (as Dummett has suggested) a "norm of assertability" or, as Frege says, that it "points the way" for logic, where logic is, again, the theory that articulates the *laws* of truth rather than simply those of what is held true.[54] Any such theory will never, if Frege's argument is correct, amount to a general *definition* of truth.[55] But even so, it nevertheless remains open to hold that, as Frege himself says, the *phenomenon* of truth is "indicated" or "pointed toward" by the logical/semantic theory that articulates the logical structure of any possible language. In this sense, a logical/semantic theory that articulates the constitutive logical structure of predication gestures beyond the properties and relations of beings, toward the constitutive and systematic connection of the sentences of a language to their truth-conditions. Beyond beings, such a theory would thus gesture toward the indefinable point at which the sentences of a language bear, in the structure of predication itself, the inarticulate mark of their being.

As is well known, Frege officially distinguishes "assertoric force," the particular kind of force that is applied to a sentence or its content when it is asserted, from the content itself. Along with this, he sometimes holds that *asserting* is just one of several possible actions that can be performed with one and the same content (including also, for example, questioning, denying, merely considering, etc.). In a 1906 summary of his "logical doctrines," for instance, Frege placed at the top of the list his having "dissociated assertoric force from the predicate."[56] Along similar lines, he treats one component of what would become the familiar logical "turnstile" as a "judgment stroke" signifying that what follows is in fact judged, or that its truth is recognized.[57] According to Frege, it is necessary in particular to use the symbolism, because it is necessary to separate the act of judgment, when it occurs, from its subject matter; for otherwise it would be impossible to distinguish the judgment (or assertion) that some content does in fact hold true from a mere supposition of the same content.[58] The maintenance of both claims (about the separation of assertoric force from the content and about the significance of the judgment stroke) invite difficult questions about the proper description of the activity of judgment, and of force itself, from a Fregean perspective.[59] One problem here is that of how the logical structure that characterizes a language and determines the sense of its terms constrains or is constrained by the dynamic actuality of its use, especially insofar as this use characteristically involves the possibility of making assertions that are true or false. More generally specified, the problem is that of the general (linguistic or nonlinguistic?) context in which *force* communicates with *meaning* in general, or the broader contextual space *in which* the specific structure of communicative sense relates to that of meaningful intersubjective action and *praxis*.[60]

Both Heidegger and Frege, in different but complementary ways, thus point to a transformative deepening of the problematic that is already laid out in the Visitor's discussion of the positions of the materialists and the friends of the forms. The problematic is that of the *relationship of thought to being* that allows, within what is there determined as a human life, for the *possibility of thinking being as such*. If the solution suggested by the Visitor to the problem of this relationship has set the terms for the subsequent development of thought about it in the Western tradition, both Heidegger and Frege, in their radical thinking about the logical structure of the sentence, challenge the specific solution by deepening the general problem to which it responds. As we shall see in more detail, Heidegger does so by pointing toward the more original *ontological* phenomenon of truth as *unconcealment*, which relates Dasein to the very structure of world as such, thereby relating Dasein's factical being back to its sense, and thereby to the being of entities. Along similar lines, Frege does so by pointing to the place of impersonal and ineffectual *sense* and to its constitutive link with indefinable truth. But if truth is indefinable for Frege, being is equally so for Heidegger, and for similar reasons.[61] In particular, in both cases, the attempt at definition collapses the phenomenon into an empirical or ontic one that appears then to be both ubiquitous and empty. This is the basis of the prejudice which, according to Heidegger, keeps us from raising the question of being today. But it is also the basis of the inevitable vicious regress which, according to Frege, results from any ontic account of the basis of predicative truth. In neither case, however, does the impossibility of definition point to the emptiness or meaninglessness of the phenomenon itself. Quite to the contrary, in fact: for both philosophers, the reinterpretation of what at first appears as the mere indefinability of being and truth as, instead, a *positive* phenomenon, grounded in the matters themselves, is the first step toward a systematic clarification of their logical and ontological sense.

One essential aspect of this clarification is Frege's development of the idea of *sense* as *mode of presentation* [*Art des Gegebenseins*]. According to this conception, the sense of a nominal term is a *mode of presentation* of a referent, and the sense of a sentence as a whole is a *thought* with a truth-value. Here, then, the sense of a term is not understood as itself anything like an entity; nor is it an internal representation or an activity of producing or combining representations. Rather, it is understood in connection with both *truth* and the presence of what *is presented* to the mind as their conditions of possibility, the logical/ontological basis *for* the presentation of whatever comes to presence. In arguing for this conception and the consequent distinction of sense from reference, Frege most centrally appeals to the need to account for the possibility that a true *judgment of identity* has the value of positive knowledge.[62] As Frege recognizes, it is not sufficient, in accounting for this possibility, simply to assume (as on his own earlier account in the *Begriffsschrift*) that the informativeness of an identity judgment consists in its recognition that two *signs* designate the same object. In order to explain this value, it is thus necessary to recognize that a positive identity judgment concerns not simply the signs, but rather the different *modes of presentation* of the object itself.

The structure of these ways or modes of givenness is spelled out in the concept of objective sense, which accordingly cannot be identified with any ontic domain of entities but rather exhibits the constitutive link between truth and their presentation. It is on this basis that Frege accordingly suggests that the *reference* of a true or false sentence is one of the two unique objects, "the true" or "the false": just as the sense of a nominal term determines the object to which it refers, the sense of a sentence determines a truth-value, which must then be considered to play the role of its referent. Though this does not resolve the question of the ontological status of these "objects," we can nevertheless understand the underlying suggestion here, again in the Heideggerian jargon, as pointing to the thought that the employment of a true sentence, in addition to presenting the referent or referents of its nominal terms, *also* structurally presents or "codiscloses" the *specific phenomenon* of truth itself. That this structure is manifest in the structural form of the sentence, again, indicates the privileged logical connection between that structural form and (to use one of Heidegger's expressions for the fundamental topic of his inquiry into logical structure) "being in the sense of truth." The structure thereby elicited, and articulated (without being defined) through the articulation of the laws of logic, is the one that Frege identifies in the *Grundgesetze* passage as the objective and nonactual domain of the contents of judgments, ultimately demanded by the objectivity and impersonality of thought itself.

The ontological implications of this link of sense and truth, beyond any identification with entities, to their presentation or presence can be further clarified through a consideration of Frege's own most important innovations on the level of formal-logical syntax. These are the new devices of quantification which he first developed in the *Begriffsschrift* and applied there to consider the logical structure of mathematical judgments.[63] Familiarly, these devices allowed him to repudiate and replace the subject/predicate logic that had been widely presupposed since Aristotle.[64] On Frege's new conception, it is generally impossible to portray the actual logical structure of a predicative sentence simply by means of the grammatical distinction between subject and predicate. Rather, within the sentence it is necessary to distinguish concept-words from object-words, and to recognize the fundamental difference in their logical nature. Whereas an object-word has an object as its reference, what a concept-word signifies is inherently "unsaturated," or requires completion by means of an object. More generally, the references of concept-terms are *functions* from objects (or sequences of objects) to truth-values. This leads to a fundamental distinction in the modes of signification of concept- and object-words, which issues in the notorious restriction on which Frege rigorously insists: namely, that a concept-word must never be used in the logical place of an object-word.[65] It is thus impossible to *refer* to concepts as we would to things. Similarly, we cannot predicate *of* concepts; as much as we would like to say "the concept 'horse' is a concept easily attained," to do so would be to violate a fundamental aspect of logical structure and to utter what could only be nonsense.[66]

As a consequence, the reality or event of predication can never be understood, on this picture, as simply forging an *ontic* relation (that is, one between fully "saturated" entities). Similarly, for Heidegger, the being of beings, such as it is expressed or indicated in a predicative sentence, is never itself a being. Or equivalently: *between* what metaphysics since Aristotle has thought of as substances and what are thought as their properties, there lies the genuinely *ontological* structure of unsaturated concepts and saturated objects, which links sentences in a language to the conditions of their truth. Like the ontological difference itself, the distinction between concept and object thus points, in a fashion that resists direct summarization and also leads to its own paradoxes of meaning, to the generation of sense at the ontological point of the insistence of a fundamental difference that cannot be schematized in a figure or captured as a substrate.[67]

Only a superficial reading could identify Frege's conception with a "Platonism" in the sense in which that term is often used today, that in which it refers to a "theory of the forms" as substantial, timelessly existing entities connected to ordinary sensory objects by an obscure relationship of "participation." The distinction between concepts and objects, for example, by itself suffices to clarify that the unity of a predicative sentence on Frege's conception is not founded in any relationship between two self-standing (or "saturated") entities of any kind. Rather, it is founded in the peculiar kind of unification that occurs in the figuring of a specific object as a definite value of a variable function.[68] In the further development of this conception of unity made by Tarski, the formal theory of truth turns on the interpretation of this relation as "satisfaction." The dynamics of this relation are modeled, not by any mimetic or representational account of the relation of universals to individuals, but rather by the mathematical set theory developed by Cantor and Frege, among others, at the inauguration of the analytic tradition.

Does Frege's distinction between concept and object, together with the radically new picture of logical structure and sense that it involves, then, suffice to *solve* the ontological-temporal problem that Plato's Visitor poses in the *Sophist*? It does not, for at least three interrelated reasons. First, since both concept-terms and object-terms have reference for Frege, his invocation of concepts as "unsaturated" appears to require *ontologically* the existence of unsaturated or "gappy" entities—objects of reference that inherently contain "missing" places or have undetermined aspects corresponding to the variable place or places in a propositional function. But it is mysterious what such entities could be, how they could enter into relationships with ordinary (nongappy) entities, or what these relationships could be. In drawing the *logical* distinction between concept and object, Frege thus leaves the *ontological* status of this distinction radically unclear.

Second, there is the deep and transformative problem on which Frege's attempt to ground arithmetic in logical rules ultimately foundered, and to which we will return in subsequent chapters: that of Russell's paradox of the set of all sets that are not members of themselves. The paradox results on the assumption Frege made in formulating his basic law V in the *Grundgesetze*: that of an unlimited comprehension principle, or an unrestricted correspondence between intensions and extensions.[69] Ontologically viewed, this is just the assumption of an unlimited correspondence between what is conceptually *thinkable* as the characteristics, properties, and relationships of entities, on one hand, and entities themselves determined as *having* these characteristics or properties (or standing in those relationships), on the other. But when applied to the consideration of the ontic *totality* of entities as an existing whole, it yields the contradiction witnessed in Russell's paradox, that of a totality that both is and is not self-membered.

Third and perhaps most indicatively, Frege's distinction between concepts and objects is itself founded in the threefold distinction among propositional *functions*, the objects that can "fill" their variable places, and the *truth-values* of the sentences thereby generated. It thereby understands concepts and conceptual functions themselves as semantic *rules* and suggests the understanding of a language (whether natural or artificial) according to which its semantic structure is wholly governed and exhausted by the totality of its rules of this kind. This understanding of concepts as semantically determinative rules has far-flung consequences for the subsequent development of thinking in the analytic tradition about formal and natural languages. But it also—and already in Frege's original formulation—raises the ontological question of the mode of existence and availability of such rules themselves. In particular, there arise here the questions of the *institution* and *origin* of semantic rules, their communication and learning in the process of acquiring or interpreting a natural language, and above all the basic ontological relationship between a (conceptual) rule and its determinate instances of application: the (in general, infinitely) many entities, objects, or relationships to which it applies (or which, when substituted in the requisite variable "places," yield a *true* sentence).

These three problems appear at first glance to concern different aspects of Frege's philosophy, but if viewed in the context of an ontological/temporal investigation grounded in the question of ontological difference, they are in essence the same. All three concern essentially the *ontological* status of concepts: their peculiar mode of being or existence and the relationship of this mode to that of objects or entities more generally. All three, moreover, raise this question of the ontological status of concepts in its codeterminative relation to the structure of (propositional) *truth*: that is, specifically, the structure by which conceptual or propositional functions are determined, in application to specific entities and relations, as producing sentences, propositions, or contents capable of being true or false. As such, they raise the further question of the specific kind of *temporality* of what Frege understands as concepts and senses: how they are themselves temporally situated—if at all—and how then to understand their relationship to the objects and truth-values they present in the diachronic activity of thought and language.

In one of his last works, Frege notoriously attempts to specify the specific ontological mode of senses or contents in terms of the metaphor of their inhabitance of a mysterious "third realm," separate both from spatiotemporal objectivity and from subjective experience, and independent of the temporality of both.[70] But as is evident in the inherent obscurity of the metaphor of "grasping" by means of which Frege characterizes our relationship to contents residing in this realm, this is by no means a solution to the Platonic problem of the temporality of thought in its relationship to logical/categorical structure. Rather, it is a restatement of it, on (what remain) deeply unclear ontological/temporal foundations. In particular, the obscure metaphors of the timeless third realm and our (temporally situated) grasping of content give expression to the same problem that the Visitor originally poses to both the materialist and the friend of the forms at the outset of the determinative battle of gods and giants in the *Sophist*: that of the *relationship* between temporally situated thought and knowledge, on the one hand, and the (apparent) atemporality of what is true or real in itself, on the other. But although Frege thus does not solve the Visitor's problem of truth and time, his thinking about the ontological implications of formal logic nevertheless does provide, as I shall argue in more detail in subsequent chapters, the essential conditions under which this problem can be brought out and posed today. Specifically, this thinking provides the conditions under which it can again be posed as the problem of the underlying relationship of thought to presence.

4. Heidegger: From Onto-theo-logy to Philosophical Chronology

In the 1925–26 course *Logic: The Question of Truth,* given just one year after the *Sophist* course, Heidegger considers the specific possibility of the false *logos* through a close analysis of Aristotle's *Metaphysics,* book 9, chapter 10.[71] In the chapter, Aristotle discusses the meaning of truth and falsity with respect to both composite and incomposite things. Whereas, with respect to composite things, being amounts to synthetic unity and the truth of statements about them thus depends on whether or not the composite is unified (Aristotle gives the examples: "The wood is white" and "The diagonal is incommensurable"), in the case of noncomposite or simple entities or statements about them, neither being nor truth can be understood in terms of *synthesis* and *diaeresis.* In fact, according to Aristotle, in these cases there is *no* possibility of falsehood; all that is possible to do is to "simply touch and address" the entity. By contrast, the possibility of falsehood, error, or deception, where it does exist, is then dependent, according to Aristotle as Heidegger reads him, on the possibility of a synthesis of these ultimate entities into a *unity of copresence.* This possibility of synthesis is itself, in turn, dependent on the *standing existence* of the ultimate noncomposite elements that are themselves *simply* and *always* present.[72]

The analysis confirms, according to Heidegger, that Aristotle understands being itself as "presenting or presence [*Präsenz, Anwesenheit*]" and understands the "corresponding comportment [*Verhalten*] to beings as beings" as one with the basically "presentative [*Präsentisch*]" character of being the "rendering present [*Präsentierens*]" or "making present [*Gegenwärtigens*]" of something.[73] Whether it is a case of the synthesis of composite entities or the simple encounter with things that always are, this "making-present" is furthermore, according to Heidegger, a *temporal* phenomenon. In particular, it is "letting a present being be encountered in a now-moment [*Anwesendes in eine Gegenwart begegnen lassen*]," and this "'presence-now is a characteristic of time [*Gegenwart ist ein Charakter der Zeit*].'"[74]

As Heidegger suggests, Aristotle's conception of the synthetic structure of the *logos* that makes falsehood possible is closely related to Plato's conception of the logical basis of the unity of the sentence, as articulated in the *Sophist.*[75] On both conceptions, in particular, the synthetic unity of the sentence is, on the one hand, the logical unity of subject and predicate, and on the other, the ontological synthesis of basic and eternally subsistent elements. But if predicative *synthesis* thus is, as Heidegger says, for both Plato and Aristotle a "chameleon-like concept [*schillernder Begriff*], sometimes logical, sometimes ontological—or more precisely, usually both at the same time," then the critical question of the temporality of the *logos* requires that this synthetic conception *itself* be further interrogated in terms of the conception of temporality that makes *it* possible.[76] This conception is, as Heidegger argues, none other than the conception of presence as grounded in the unity of a "now," or in the simultaneous copresence that allows the articulate structure of the *eide* to appear together in the unified nexus of the sentence and in the unified cognitive act of the thinking soul.

Once the temporal question is posed, it is no longer possible, Heidegger suggests, to see the structure of the *logos* as it is involved in making linguistic statements *simply* as a matter of synthesis at all.[77] Rather, it is necessary to undertake a deeper analysis of the specific way in which the structure of the *logos,* and with it truth or falsehood, is bound up (in what Heidegger admits is still an "enigmatic" way) with an underlying structure of time. The task here suggested is to "use our insight into the inner connection between truth—or uncoveredness as presence-now—and being as presence" to elicit a more fundamental connection between truth and time.[78] As a guideline for this task, Heidegger proposes the thesis that "truth, being, and consequently falsehood, synthesis, and assertion are in some kind of, up until now, wholly obscure sense, connected [*im Zusammenhang*] with the phenomenon of time," and proposes to use the specific conclusions of the analysis of Aristotle's account of falsehood as a basis for demonstrating these connections.[79] This points, in particular, to the project of a "phenomenological chronology" which will demonstrate how the phenomena of assertion, truth, and *logos* are "essentially" characterized by time and rooted in its underlying structure.

Near the beginning of the summer 1925 course titled "History of the Concept of Time," Heidegger further specifies the task of such a phenomenological chronology as that of a "history of the discovery of time [*Geschichte der Entdeckung der Zeit*]" and a "history of its conceptual interpretation [*begrifflichen Interpretation*]." This history amounts to a "history of the question of the being of entities" and in particular of "the attempt to uncover [*entdecken*] entities in their being."[80] This investigation will take as its guideline the particular understanding of time that has made it possible to determine the various domains of reality according to their temporal characteristics, for instance as "temporal, extratemporal," or "supratemporal reality."[81] The phenomenological chronology thus investigates the temporal conditions for the separation of domains of entities, most significantly the separation of the domains of "nature" and "history," and provides the possibility of indicating the "original and undivided context of subject matter [*Sachzusammenhang*]" that "remains hidden" behind the division by illuminating the specifically temporal basis of the concepts of positive science and their origin in pretheoretical experience.[82]

In the 1928 course *The Metaphysical Foundations of Logic*, Heidegger points to the particular way in which such a chronological investigation of traditional logic deepens the problematic of the nature of "reason" in the sense of "ground." In particular, according to Heidegger's reading in the course, Leibniz, along with the tradition since Plato, understands the specific structure of the *logos* as that of a *synthesis*. Here, a judgment is "in the broadest formal sense" a "relationship between representations" or "of concepts."[83] For Leibniz, this is essentially a structure of inclusion: in a true judgment, in particular, the predicate or its concept is "in some way" involved or included in the concept of the subject.[84] The consideration, according to Heidegger, wavers between an ontological level, on which the containment is the real containment (or *Inesse*) of a property in a substance, and a purely logical one, on which the primary relation of containment is exhibited in the true sentence or judgment itself.[85] For both determinations, the specific nature of the connection is furthermore (insofar as truth is understood as *idem esse* or *identity*) here thought as the basis for the unity of the unitary structure of the sentence or judgment as such. On Heidegger's reading, this conception in turn provides the basis for Leibniz's own conception of the monad as individuated drive, capability of representative apprehension and reflection of the totality of the world.[86] It further underwrites, Heidegger suggests, Kant's subsequent determination of the highest principle of subjective unity as that of the unifying synthesis of transcendental apperception.[87]

Leibniz's notorious principle of sufficient reason expresses the implication of this originally and undecidably logical/ontological unity for the problem of *grounding*. This is the problem of the nature of beings, understood ontologically and metaphysically as the problem of the *basis* of beings as a whole. In this form, the problem of grounding is, according to Heidegger, specifically that of a "rather than" which takes up being insofar as it asks: why beings *rather than* nothing?[88] If this question of grounding is radicalized along the lines suggested by invoking the problematic of the ontological difference between being and beings, the unity of the monad as subject is replaced, Heidegger suggests, with the "transcendence" of Dasein, understood in terms of its constitutive possibility of disclosing its world. Such a relationship is not simply a relation of Dasein to an (ontic) totality of entities. Nevertheless, in a suggestive "appendix" to the first section of the second part of the course, Heidegger points to the need to develop, alongside or before the fundamental ontology of Dasein, a "metontology" capable of considering "nature" as a possible and thinkable totality of beings as a whole within which Dasein *itself* is an actual existent:

> Since being is there only insofar as beings are already there [*indem auch schon gerade Seiendes im Da ist*], fundamental ontology has in it the latent tendency toward an original metaphysical transformation which becomes possible only when being is understood in its whole problematic . . . In other words, the possibility that being is there in the understanding [*das es Sein im Verstehen gibt*] has as a presupposition the factical existence [*Existenz*] of Dasein, and this in turn presupposes the factical extantness [*Vorhandsein*] of nature . . . From this there results the necessity of a special problematic which has as its theme beings as a whole. This new investigation resides in the essence of ontology itself and is the result of its overturning [*Umschlag*], its *metabole*. I designate this problematic: *metontology*.[89]

This problem of Dasein's possible reflection of the world from a position within it is further manifest, according to Heidegger, in the structure of "world-entry [*Welteingang*]" in which both particular entities, and the more general domains in which they exist and are definable, become accessible and possibly intelligible to Dasein.[90] As grounded in the basically *temporal* phenomenon of the constitution of a "horizon of possibility in general" which subsequently allows any definite possibility to be "expected," both Dasein's structural transcendence and the phenomenon of world-entry it allows are themselves based, according to Heidegger, on a more "primal" "temporalization of temporality" in which underlying temporality constitutes itself in a particular way.[91] Both the structure of Dasein's transcendence and that of the underlying (self-)temporalization of basic temporality, remain, however, in a questionable relationship to the phenomenon of the world understood as the *totality* of beings:

> Time is essentially a self-opening and expanding [*Ent-spannen*] into a world. I will not go into the comparison [with Leibniz] any further, particularly the question of the extent to which one might conceive the interpretation of Dasein as temporality in a universal-ontological way—just as the monadology is presented as an exposition of the whole universe of beings. This is a question which I myself am not able to decide, one which is completely unclear to me.[92]

What is literally undecidable for Heidegger in 1928 is thus the question of the relation of Dasein's being in the world to the ontic *totality* of beings *and thus* to the temporality at the basis of the possibility of totalizing them, or of *quantifying* over them as a whole. As I shall argue in subsequent chapters, this *undecidability* of the question of world and world-entry in relation to the ontic totality of beings, which remains characteristic of Heidegger's inquiry into the conditions for the historical intelligibility of beings through its whole itinerary, can also be understood as a positive and constitutive phenomenon, and thereby recovered for phenomenological research. If, in particular, the problem of world-entry is always a problem of the constitution of the (infinite) sense of the whole of beings *for* a (finite) Dasein, the undecidability indicated here is intimately connected to the problem of reflection and totality that emerges from the aporeatic relationship of finitude and the ("mathematical") infinite themselves.[93] This is the problem of how an "infinite" structure—including eminently the kind of structure that characterizes the "sense" of beings as such—emerges and is thought at a particular time by finite beings as bearing on the determination of the whole of what is.

2

The Early Heidegger and the Givenness of Form

Between 1919 and 1926, Heidegger developed the main elements of the methodology applied in the "preliminary" ontological analyses of Dasein and of Dasein's relationship to time that comprise the finished and published portion of *Being and Time*. As I shall argue here, the methodological devices Heidegger developed over these years also have a twofold significance for the current inquiry into the relationship of truth and time. First, they show how a genuinely *hermeneutic* methodology can illuminate formal and ontologically constitutive structures of both phenomena, clarifying their sense and their relationship to the sense of being by indicating the form of their givenness in factical life. Especially important in demonstrating the possibility of this clarification is, as we shall see, the methodology of *formal indication*, which points to and articulates the formal underlying structure of what is in each case given, in its concreteness, in the singularity of a concrete life. Second, they show how a hermeneutics of facticity in this sense can reopen the decisive question of the structure of *given time*: that is, of time as experienced, thought, and measured by an entity, Dasein, which is itself defined by the ontological difference between being and beings, and by its own structural and lived involvement in this difference.

1. Husserl: Categorial Intuition and Transcendental Logic

In a 1963 retrospective essay devoted to tracing his own "way to phenomenology," Heidegger recalls one of the particular topics in Husserl's phenomenology that, as he studied with Husserl in the early 1920s, yielded a special directive for Heidegger's own developing sense of the application of phenomenological methods to the problem of being.[1] This is Husserl's theory of *categorial intuition*, or the direct intuition of abstract and categorial structures. Husserl had developed it in the sixth investigation of the *Logical Investigations*, there sharply distinguishing it from more familiar forms of sensuous or sensory intuition. As Husserl already suggests in the *Investigations*, the idea of categorial intuition is crucial for a phenomenological account of the cognitive availability of formal/logical categories, and thus for the possibility of understanding logical form. By drawing on and transforming Husserl's concept of categorial intuition in the altered context of the question of being, Heidegger is subsequently able, as we shall see, to develop his own account of the availability of ontological structures and features to a reflective phenomenological inquiry. This clarifies the possible form of a logical, ontological, and ultimately temporal investigation of the formal structure of the being of beings, and of being in itself.

Most directly, Husserl's theory of categorial intuition arises in response to questions internal to his own phenomenological account of truth and knowledge. On this account, knowledge consists in a particular kind of "identifying" synthesis between an intentional act that intends an object or state of affairs and a distinct act of intuition or presentation that "fulfills" this intention. For instance, in a simple perceptual act, an object of perception may be intended in an unfulfilled way as I take myself to see it; if the object indeed exists and presents itself, the act is fulfilled and knowledge is attained. Husserl understands this as an "identifying act" that connects the meaning-intention with the actual presence of the object as it is in itself and is directly given.[2] Indeed, Husserl suggests in the introduction to the sixth *Logical Investigation*, it is in terms of such a "synthesis of fulfillment" that we must understand the phenomenological idea of "being in the sense of truth." This synthesis characterizes, Husserl says, the "single . . . phenomenological situation" on which all of the "varying notions of truth" must be based. The complete fulfillment that occurs in the "identifying synthesis," moreover, is, Husserl says, "'correspondence' rightly understood, the *adequatio rei ac intellectus*."[3]

It is this conception of the universal structure of truth that leads Husserl, in connection with the complexities of the various kinds of intentional acts, to introduce the concept of "categorial"—as distinct from merely "sensuous"—intuition. In a simple act of perceiving an object, knowledge can be understood as the correspondence of an intending act with a purely sensory fulfillment that presents that individual object in its direct appearance. Here, for instance, I perceive a red surface; the sensory quality of redness is given directly and simply in the fulfilling intuition. But it is very often the case (as, for instance, in the acts underlying even simple *predicative* judgments and assertions) that what we have knowledge of has, essentially, a form and structure that goes beyond simple, individual perceptual presentations. I may have, for instance, a perceptual experience that yields not only the simple seeing of an individual object, but also of *how things stand* with the presented object (or objects) as well. For instance, in a statement based on direct, intuitive perception, it is possible not only to claim that one sees individual objects (for instance, "this paper"; "an inkpot") but also that one sees *that* various states of affairs obtain (for instance, "that the paper has been written on, that there is a bronze inkpot standing here," etc.).[4]

This possibility, that complex and structured states of affairs as well as simple objects can be given intuitively, complicates the phenomenological account of truth as "identifying synthesis." Accordingly, we must pose, as Husserl now does, the question "What may and can furnish fulfillment for those aspects of meaning which make up propositional form as such, the aspects of '*categorial form*' to which, e.g., the copula belongs?"[5] Since, Husserl says, there is nothing in the sensory givenness of the objects to correspond with elements of the complex judgment that are expressed by words such as "'the,' 'a,' 'some,' 'many,' 'few,' 'two,' 'is,' 'not,' 'which,' 'and,' 'or,'" and so on,[6] we must recognize, in addition to sensuous intuitions, the possibility of a distinct kind of wholly nonsensuous intuition, which Husserl terms "categorial." This kind of intuition yields knowledge of the possible forms of objects and their combination and relation, including the sense of "being" as expressed in the copula of a predicative assertion such as "The paper is white."

More generally, Husserl says, all talk of "logical form" must be understood in terms of the kind of structure that is added by categorial intuition to the simple "material" of objects presented.[7] In any case of complex, structured presentation, there will be aspects of what is presented that go beyond the simple, sensory material itself. In each such case, Husserl argues, it is therefore necessary to acknowledge the givenness of structure or form that goes beyond mere sensory intuition, yielding possible knowledge of the properties, aspects, and relations (logical as well as empirical) of the things as they are in themselves. Indeed, according to Husserl, it is in categorial intuition that *universals* and types are first given to us as objects of knowledge. In particular, on the basis of several acts of intuition of individuals, it is possible that a new kind of "abstractive" act occurs in which an object of a new type is given: the universal under which each of the several particulars stands.[8] Although this new objectivity remains abstractively *founded* on wholly sensory presentations, it also goes beyond the individual sensory presentations in yielding intuitive access to an "ideal object," the "very sort [*diese Art selbst*]" shared by them.[9]

In his *Formal and Transcendental Logic* of 1929, Husserl further clarifies this phenomenologically crucial dimension of the *logical givenness of form* as supported by the possibility of categorial intuition. According to

Husserl, the study of the possible relations of judgments in accordance with fundamental laws of validity, and most importantly the law of noncontradiction, yields a "formal logic" consisting in a "pure analytic apophantics" or a general theory of the possible forms of judgment and possibilities for their conjunction in the unity of a statement or theory.[10] This is to be supplemented, though, with a "transcendental logic" that, going beyond the mere constraint of the law of noncontradiction, yields a synthetic *a priori* theory of the givenness of material provinces, or regions of entities, "in one totality," that is, the totality and unity of "world."[11] In accordance with the later Husserl's transcendental idealism, the domain of transcendental logic is seen as constituted by a transcendental subjectivity that gives unity and normativity to all possible empirical theories of the world, including any account of the world as a whole. Nevertheless, this "transcendental" logic remains rigorously grounded in the first, purely analytic and apophantic level of logic, and in a certain sense both actually remain "formally" determined. In particular, both levels of logical theory remain wholly dependent on nonsensory, categorial intuition of form to specify their relative articulations of the specific formal and material ontological domains that they yield. In both cases, what remains essential to this is the possible givenness of *ideal* structures and objects. In this givenness, according to Husserl, a consciousness is formed of the ideal judgment, argument, and so on, as "not merely quite alike or similar" in various empirical instances, but rather "numerically, identically, the same."[12] Thus, although the instances of the "appearance of the judgment" in consciousness may be multiple, it is possible to attain an intuitive awareness of the judgment itself as an objectivity outside of temporal determination and identically the same in each of its appearances.[13]

2. Dasein and the Hermeneutics of Facticity

From the beginning of his academic career, Heidegger's own investigations of contemporary thinkers and the historical tradition show a deep concern with the question of the source and provenance of *categories* in relation to the psychology and phenomenology of the thinking subject. In the first, doctoral qualifying dissertation of 1913, *The Doctrine of Judgment in Psychologism,* Heidegger applies Husserl's critique of psychologism about the basis of judgment to the theories of Wilhelm Wundt, Theodor Lipps, Heinrich Meier, and Franz Brentano. Two years later, the *Habili-*

tationsschrift, titled "The Theory of the Categories and Meaning in Duns Scotus," had exhaustively examined the problem of the origin of categories in Scotus's scholastic philosophy, with particular emphasis on the origination of a "material determination" of formal categories in relation to the phenomena of unity, reflexivity, and what Scotus called *haecceitas* or "thisness."[14] Here, Heidegger already draws also on the phenomenology of Emil Lask, who had developed Husserl's doctrine of categorial intuition to describe the pretheoretical or precognitive givenness of categories in immediate, factical experience.[15]

In the course "The Idea of Philosophy and the Problem of Worldview," given in the "war emergency semester" of 1919, Heidegger takes up the problem of the "essence of worldview" and considers critically the methodological and thematic views of the neo-Kantians Natorp, Rickert, and Windelband about our access to categories, the forms of judgment, and the sources of value and truth. For the neo-Kantian "philosophy of value," truth appears only in the category of what "holds value [*gelten*]" or what amounts to a teleological norm for thought. Here, the category of "value" in this sense is moreover sharply distinguished from the existence of what *is* in being: the realm of what holds value and truth is conceived as wholly distinct from actual existence, and the relationship between the realms is left somewhat obscure. But, Heidegger objects, it would be impossible to determine the normativity of value and truth, or what is to be valued or taken as true, if this "normativity" did not somehow have a basis in what is given in the concreteness of actual, lived experience.[16] Accordingly, Heidegger next attempts to determine the possibility of the "material giving" by which concrete, pretheoretical life experience yields access to the theoretical categories by means of which we can understand the structure of judgment and the possibility of truth. Here, what is thought in neo-Kantianism as the "psychic" realm of the givenness of ideals and values must be interrogated as to its own status and kind of existence: thus the whole problematic "concentrates itself" on the single question of "how the psychic is to be given as a sphere."[17] And since, according to Heidegger, "the theoretical itself and as such refers back to something pre-theoretical," it is essential to determine how the structure of categories might already be given in pretheoretical experience and in the kind of availability of objects that is displayed in ordinary, nontheoretical life.[18] This demands, in particular, that we come to see concrete lived experience as a continuous temporal flow of change already endowed with meaning, not what "pass[es] in front of me like a thing" or even "a fixed process," but rather an "ongoing event" which is neither "inner" nor "outer," neither "physical" nor "psychical," but rather "lives out of one's 'own-ness'" and only in this way.[19]

How, then, does the flux of concrete experience, with its event-like, appropriative character, actually suffice to point toward the formal categories of judgment, being, and truth? And how might these categories actually be determined theoretically on the basis of their pregivenness in concrete, lived experience? Heidegger sees Husserl's phenomenology, here understood as a "pre-theoretical primordial science," as holding the answer to these problems.[20] In particular, a rigorous practice of "phenomenological seeing" allows the "pre-worldly" "experienceable as such" to be elicited, whereby it points toward the level of the "formally objective," which is not limited to the categories and structures of objects but instead refers back to the "fundamental level of life in and for itself."[21] The key to this reciprocal movement between pretheoretic experience and theoretical recapitulation is what Heidegger calls, in the closing pages of the course, a "hermeneutical intuition" that functions as an "originary phenomenological back-and-forth formation of the recepts and precepts from which all theoretical objectification, indeed every transcendent positing, falls out."[22]

In the 1923 course "Ontology—The Hermeneutics of Facticity," Heidegger further specifies the "hermeneutical" methodology of the eliciting and demonstration of the formal dimension of ontology from out of the pretheoretical factical givenness of concretely experienced life. Emboldened by his decisive and transformative confrontation with Aristotle, Heidegger is now willing to criticize the totality of "traditional ontology" since the Greeks, as constantly having in view as the basis for its interpretation of being what is really only a specific modality of it, namely "being-an-object." In so doing, Heidegger suggests, traditional ontology systematically blocks access to "that being [*Seienden*] which is decisive within philosophical problems," namely Dasein itself.[23] In particular, Heidegger says in the course of a highly critical discussion of Max Scheler, the ontological tradition presupposes the guideline of the traditional definition of man as the rational animal, and thereby places extant, objective beings in advance in "definite categorial forms" that are predetermined by this definition.[24] By contrast, hermeneutics in Heidegger's sense has as its thematic object "in each case our own Dasein in its being-there for a while at the particular time [*jeweils unser eigenes Dasein*]." Heidegger defines "facticity" in terms of the "ownness" and "our own [*eigenen, Eigenheit*]" of this being.[25] He emphasizes that Dasein, so conceived, is not to be understood as "human being" in any familiar sense, or indeed as an answer to the question "What is man?" Relatedly, facticity as the Dasein which is in each case our own "initially contains nothing of the ideas of 'ego,' person, ego-pole, center of acts."[26] Indeed, in the hermeneutics of facticity, the expressions "human Dasein [*menschliches Dasein*]," "human being [*Menschsein*]," and "being of

Man [*Sein des Menschen*]" are explicitly to be avoided.[27] Further, "even the concept of the self is, when employed here, not to be taken as something having its origin in an 'ego' [*nicht 'ichlichen' Ursprings*]!"[28]

The critique of the humanism of the rational animal, which Heidegger voices here, will remain, in various forms, a fixture of his critical discussions of the tradition throughout much of the rest of his career. Here, although Heidegger still conceives of his own method of hermeneutical inquiry into facticity as simply an application of Husserl's own phenomenological method of seeing, the critique nevertheless yields the terms in which he will first begin to offer some cautious criticisms of Husserl's project, or at least of what he sees as its "misunderstanding" in the further development of phenomenology, in Husserl and others, after the *Logical Investigations.* On Heidegger's gloss, phenomenology arises in close connection with the rapid development of psychology at the end of the nineteenth century, at a time when "the work of philosophy was . . . applied mainly to the phenomenon of consciousness" and epistemology and logic were widely thought to have a psychological foundation.[29] By contrast, the *Logical Investigations* boldly asked about the kind of being possessed by the "objects about which logic speaks." This questioning yields Husserl's detailed study of content and meaning, as well as his decisive development of the concept of "intentionality" already suggested by Brentano. Nevertheless, Heidegger suggests, the *Logical Investigations* have largely been misunderstood, for instance as primarily a contribution to epistemology in line with neo-Kantianism, and the further development of phenomenology has exhibited various moments that jointly tend to distort its original sense and even render impossible its fundamental mode of investigation.[30] Accordingly, he calls for a revitalization of phenomenology that does not define it in terms of any privileged domain of being or fixed methodological model, but rather as "a *how of research* [*ein Wie der Forschung*] which makes the objects in question present in intuition and discusses them only to the extent that they are there in such intuition [*soweit sie anschaulich da sind*]."[31] This provides the essential guideline for the "hermeneutics of facticity" which always interprets the phenomena on the basis of Dasein's "fore-having [*Vorhabe*]" of them as they are actually given in concrete, factical life and in the phenomena of Dasein's "having-itself-there [*Sichselbstdahaben*]" and "always-being-in-such-a-manner [*Immersoseins*]."[32] In particular, Heidegger suggests: "The forehaving in which Dasein (in each case our own Dasein in its being-there for a while at the particular time) stands for this investigation can be expressed in a 'formal indication' [*formaler Anzeige*]: *the being-there of Dasein (factical life) is being in a world* [*Dasein (faktisches Leben) ist Sein in einer Welt*]."[33]

3. The Formal Indication of Being and Time

In thus specifying the underlying structures of Dasein and world that are hermeneutically articulated on the basis of the fore-having characteristic of concrete life, Heidegger makes use of a methodological device, first introduced in the course "Introduction to the Phenomenology of Religion" of 1920–21, which plays an essential role in his method of analysis throughout the early 1920s as well as in *Being and Time*.[34] This is the device of "formal indication [*formale Anzeige*]." By means of formal indication, the concreteness of factical experience points "back" to the more formal structures that are its basis in the structurally articulated phenomenon of Dasein. The possibility of such an indication proves decisive for hermeneutic investigation into Dasein, since it is able to point to these structures as they are in fact *already* given in concrete and singular life. In doing so, it makes explicit what is, there, typically only implicit, and thereby brings to light the actual grounding of these essential structures of Dasein as such in its own essentially ontic-ontological structure. Thus understood, as I shall briefly argue, formal indication is at least one methodology by which the hermeneutic analysis can bring out the ontological grounding and underlying structure of the phenomena definitive or constitutive of Dasein's life and world, including significantly the form of given time.

In introducing "formal indication" in the 1920–21 course, Heidegger draws on a distinction that was earlier suggested by Husserl between two importantly different kinds of universalization.[35] *Generalization* is one kind: in generalization, we move from the individual objects or phenomena to their species or type, and then from more specific types or species to yet higher *genera*. Generalization is to be distinguished, however, from *formalization*: in formalization, we do not simply move from a phenomenon to the higher genus under which it falls, but rather elicit and grasp its structure and sense through a demonstration of it as it is given in the *particular* case. We can grasp the difference between the two processes, as Husserl considers it, by considering the process of a concrete demonstration with respect to a particular figure, for instance, a right triangle that is colored and has a certain size. To *generalize* about the triangle would be to relate it as exemplary of a species to various higher *genera:* for instance, it is subordinated to the types of *triangle* (in general), *colored figure, plane figure, geometric shape*, and so on. However, what is needed to perform a specific demonstration—for instance, to show that its right angle is equal to the sum of its other two angles—is not generalization. It is, rather a *formalization,* in which the *uniquely* definitive formal features of this figure are grasped and related (perhaps with the aid of a construction). To formalize in this way is thus not simply to subordinate the various features of the figure to higher types or genera, but to identify its own formally relevant features and demonstratively indicate their structure. In this way alone, according to Husserl, is it possible to gain access to the "formal-universal" aspect of the particular case that can also be described as its "logical form."[36] Such a demonstration is not merely a definition in terms of a statement of genera and specific differences, no matter how detailed. It is rather an eliciting and actual provision of the logical form of the particular case, which alone can show the actual structure which is "filled out"—Husserl says—in the *individual* and *singular* "this-here" or "tode ti."[37]

In his own development of these ideas, Heidegger retains this concept of *formalization* as the indication of the logical form that thus demonstrates the structure and sense of a particular case, including the distinctive way in which it is *given*. But he applies it, outside Husserl's own transcendental-phenomenological conception of intuitive givenness in consciousness, to the structures broadly characteristic of Dasein's concrete temporal life, as they are given and shown there.[38]

How should we then understand the sense of both *formalization* and the *formal*, as they are employed both by Husserl and Heidegger? In broadest terms, the formal has the character of the *unification* of a plurality, the character of the "one over many" which also decisively and originally characterizes the being of the "form" or "idea" for Plato. The connection Plato draws between the form of something and its *essential* features—what captures its being or "what it is"—is also importantly retained in this sense of "formalization" (especially by Heidegger). Here, the formalization of a particular object or pattern thus captures its essentially determining aspects or features, features which *may* indeed characterize other particulars individually. But to formalize is not simply to describe these individuals, or indeed (as we have seen) to subsume them under a more general category by means of commonalities. It is, rather, to locate them within a relatively unitary (and indeed universal) relational or articulative *structure*, thereby illuminating their essential identities in their underlying ontological determination.[39] Crucial to the achievement of this demonstrative illumination in each case is the actual *seeing* of the defining structures or features *in* the individual case before us. This is what makes the formalization a genuine *demonstration* (as opposed to, say, a stipulation or convention) and thereby maintains its essential link to the broader project of phenomenology, understood as the actual *showing* of the things themselves.[40]

Formalization in this sense is often akin to definition, although we should not expect the demonstrations it yields always to be capturable in the form of an abstract, general verbal statement of essentially defining features or necessary and sufficient conditions. It operates not, as we have seen, by subsuming individuals under general categories fixed or set in advance but by demonstrating the formal features and relationships of the individual case at hand that make it what it is. In this respect, the order of explanation or demonstration is also different from that of more traditional forms of definition: rather than *beginning* with a general categorical structure or abstract definitive statements and moving toward specification and individuation, we are here starting with the *individual* case that is *already* (in some sense) present before us. For this reason, the demonstratively formalizing "definition" will not always succeed (and nor should it be expected to succeed) in yielding a generally definitive statement which can itself be expressed in complete abstraction from the presence and availability of this particular case.

It is also important to distinguish the "formal," in the sense in which it is employed here, from the sense in which it is understood, since Aristotle, as opposed to the "material" or what bears "content." This Aristotelian concept of the formal, which historically yields Aristotle's first articulation of the sense of "formal logic" in the topic-neutrality and emptiness of the syllogistic forms, itself determinatively excludes reasoning about particular cases, which Aristotle instead consigns to (what is for him) the distinct faculty of *phronesis.* However, if demonstrative phenomenological formalization of concrete cases (including essentially their mode of being given or presented) is indeed possible, then this Aristotelian distinction between the "empty" forms of reasoning and anything like material "content" is itself inadequate. The demonstrative formalization must be capable of capturing the essentially determining features of the particular cases, and as such must not be understood as "pure" or "empty" in a sense that would exclude its being expressed in determinate, concrete, and "material" terms.

Heidegger's fullest discussion of the methodology of formal indication, in the Freiburg winter semester 1921–22 course "Phenomenological Interpretations of Aristotle," comes in the course of a consideration of the task of establishing a "definition" of a special sort in response to the question "What is philosophy?"[41] In connection with the special nature of philosophy as well as philosophical "objects" of investigation, it is mistaken, Heidegger argues, to impose an idea of definition which is derived from the type of definition appropriate to the objects of particular sciences and which "develops out of" categorical logic in accordance with the demand of *classifying* the entities treated by them.[42] On this "uncritical" idea of definition, to be rejected, definition is a matter of specifying the genus and specific difference of a thing.[43] This mode of definition is indeed appropriate, Heidegger says, for "quite definite regions of objects and for objects intended in one particular cognitive context [*erkenntnismäßig ganz bestimmt intendierte Gegenstände*]."[44] But with respect to philosophy it is essential to consider a more basic sense of definition as a "decision about or determination of something . . . which determination is declared, manifested, and indicated as having to be *held and believed* [*tenendum et credendum*]."[45] In particular, what is involved in phenomenological inquiry is a kind of indicating that preserves the particular way in which the object is "genuinely . . . possessed."[46] By contrast with definition in terms of genus and species, in the kind of determination involved here, "the logic of the grasp of the object, and the conceptuality of the object in its definatory determinacy in each case [*in der jeweiligen definitorischen Bestimmtheit*] must be drawn out of the mode *in which the object is originally accessible* [*wie der Gegenstand ursprünglich zugänglich wird*]."[47] What is decisive in producing the possibility of this sort of definition is the basis of this givenness in "the situation of life in which the object comes to be experienced" and the "basic intention" in which, within this situation, experience already preliminarily "aims at" this object.[48]

Under these constraints, Heidegger introduces the "formal indication" explicitly as a kind of definition that includes a "very definite bond [*ganz bestimmte Bindung*]" to the object. This bond means that, in the indication, the indicator herself stands in a "quite definite direction of approach [*einer ganz bestimmten Ansatzrichtung*]" to it. Here, Heidegger says, the "formal" character of the "formal indication" means nothing other than that the object *itself* is indicated in such a "binding" way. The "existentiell" sense of the "formal" is, explicitly, not to understood as resting in the familiar opposition of the formal to the material, or indeed in the "eidetic" in the sense of "universal generality," but rather as just this binding character of the indicated "approach" to the object on the basis of its givenness in factical life.[49] This "formal" character of the formal indication is thus, according to Heidegger, inseparable from its being an "indication"—that is, something which (as opposed to, for example, a mere description or external, conceptual definition) points *indexically* toward something that is itself "there," though at first in an inexplicit and nonconceptual fashion, in the concrete life of the indicator herself.[50] In the formal indication, the object which is thus already possessed in an "inauthentic" mode in ordinary life is recovered in a way that points toward the "temporalization" or "maturation [*Zeitigung*]" of a more authentic mode of the possession of it.[51] This taking possession is "in a radical sense" a "being" of the "temporalization" of existence.[52] What is developed and grasped on the basis of the approach is not necessarily or simply the full and complete presentation of the object, but (also) the specific way in which it is already *given* in the situation that provides access to it, including, Heidegger says, the essential "questionableness [*Fraglichkeit*]" of this givenness.[53] As applied to "oneself and life," this questionability is, Heidegger suggests, the "authentic foundation of philosophy," to be grasped and temporalized in a "radical" and "existentiell" way.[54]

Considered as a methodology, then, formal indication has several identifiable features which mark both its similarities, and differences, with more traditional modes of definition, analysis, clarification, and indication. Here are some of them:

1. Formal indication (as Heidegger uses it) is **reflexive**. It functions, primarily (though not necessarily exclusively) in the first person, to indicate features of *my* way of being, the features of the concrete "case" that I, myself, "am." This reflexivity corresponds, on the one hand, to the sense in which the phenomenon to be indicated is already "there" for me as a constitutive part in the course of my concrete life, and, on the other, to that in which it is nevertheless initially obscure in its actual ontological structure and provenance. Although the results are indeed *universal* in the sense of indicating structures of Dasein *as such*, there is no implication that they can or should be able to be understood by something that is not Dasein, or who does not have the structures indicated as (part of) *her* way of being herself. For this reason, the results of the inquiry typically have a specific bearing on the life of the investigator who *herself* undertakes it.[55]
2. Relatedly, the formal indication does not just point to the underlying phenomena themselves as *abstracta* or detached objects, but indicates as well the **way they are given** or present in **each concrete case**: how (e.g., distinctly or inexplicitly, clearly or obscurely) they are initially and already "there" in the concrete case in question.[56] This includes indicating the diverse "modes" or variations in which they can there appear, including possibly *privative* or *attenuated* ones.
3. The formal indication is (as an "indication" rather than a description or abstractive definition) irreducibly **indicative**: it has the function of *pointing* to something which is "already" *there* in some way—although generally inexplicitly at first. Its outcome is not the abstract definition of the phenomenon but its concrete demonstration as *being* there, and its recognition as such. Thus understood, the formal indication always involves, as Heidegger often emphasizes, a particular *approach*, a specific way of grasping the phenomenon concerned and holding it in view. For this reason, the functioning of a formal indication is analogous or homologous to the

linguistic functioning of (what are often called) indexicals. These are terms whose significant use on an occasion involves the indication of something which must be *present* and is thereby pointed out.[57]

4. Finally, through the other three features, the formal indication indicates the phenomenon in its **formal-ontological** character. That is, first, it articulates and points out the phenomenon in its *formal* aspect, as opposed to its "material" "filling out" in the particular case. What is at issue is not, indeed, the contingencies of such a *particular* case, but rather the indication of its formal (and, in this sense, universal) features. Typically, this amounts to pointing out an underlying *structure* of relationships within the broader and essentially *holistic* structural context of the being of Dasein. And in thus indicating an underlying and grounding structure *of* this being, it is specifically its *ontological* formal features—the features relevant to the sense or meaning of *its* particular *way of being*—that are thus identified and indicated.[58]

We can briefly see how these four features work together by considering a quick example of formal indication from the 1923 "Ontology" course. Early in the investigation of the basic character of facticity and its self-interpretation, Heidegger considers the formal, ontological, and temporal significance of what he calls "the today [*das Heute*]." Taken in its most usual significance and simply as a linguistic term, "today" is, of course, a temporal indexical. It has the function of indicating, in ordinary discourse, the temporal present: that is, the present day at the time and place of the utterance. Heidegger's focus, though, is not only on this indexical character of the term, but on how the structure of the underlying *phenomenon,* including the ontological and temporal character of the present thereby indicated, is "there" in the concrete life of Dasein. To do this, Heidegger uses a formulation of the ontological character of Dasein that has already been suggested earlier in the lecture: that *our own* Dasein is "what it is" "precisely and only in its *temporally particular 'there,'* its being *'there' for a while* [*Das eigene Dasein ist, was es ist, gerade und nur in seinem jeweiligen 'Da'*]."[59] "Da" is itself an indexical term meaning "here" or "there." As Heidegger makes clear in *Being and Time,* this indexical character is one of the reasons for his use of "Dasein" to indicate the kind of being that we are, one that definitively has a "here." The adjective *jeweilig,* including the particle *je-* (which has the general sense of "in each case," "respectively," or "at each time"), has the indexical meaning of "at that time," but also includes connotations of being-there *for a while,* that is, for a definite or indefinite *span* of time.[60] The character of "awhileness in each case" thereby indicated will later be replaced, in the analytic of

Being and Time, by Dasein's (so-called) *jemeinigkeit*—that is, its indexical/ontological character of being "mine" in each case.[61]

The scope as well as the basis of this formal indication is thus the being of Dasein as it is, in each case, "one's own"; on this basis what is to be indicated is the particular *temporal* character of this being. But the ontological/temporal analysis can be further articulated in terms of the "today." "A defining feature of the awhileness of temporal particularity is the *today*—in each case whiling, tarrying for a while, in the present, in each case our own present [*das Je-Verweilen in Gegenwart, der je eigenen*]."[62] The temporal character of Dasein's facticity is thereby articulated in terms of the way in which, in each case, its (our) temporality is one of "whiling" in the present that is (in each case) one's own. To further indicatively explicate the "today," it is essential, Heidegger says, to have reference to a *particular* [*bestimmte*] "today." Through this particular "holding" and "having" we can further bring out the ontological character of the "today" as such. It is "the present of those initial givens which are closest to us, every-one, being-with-each-other, 'our time' [*Gegenwart des Zunächst, man, Miteinandersein, 'unsere Zeit'*]."[63] Through the indication, in other words, it becomes clear that the "today" indicates that mode of conceiving of and inhabiting the present which characterizes that which is "at first" or "closest," what we take as "given." This indicates also the *temporality* on the basis of which we take ourselves *collectively* to inhabit a shared sense of these "givens," that on the basis of which we envision and act according to what "one" does. Within this temporality and the "public" character of conception and action it makes possible, Heidegger says, Dasein has a particular kind of "openness [*Öffentlichkeit*]" and a particular way of interpreting itself. This way of interpreting itself takes the form of a particular kind of discourse about itself, that of "idle talk [*Gerede*]."[64] This kind of talk displays a typical kind of "averageness" or "insensitivity to difference," marked in its structure whereby "everyone" or "one" in general [*man*] can indifferently understand it or follow it. As such, the "being-interpreted in the today" that is indicated as the formal-temporal basis of this kind of talk is thus characterized by the "averageness" in which "the innocuous initial 'givens' of the day which are closest to us" are present as "given . . . for-the-most-part and for-most-of-us."[65] As such, it is a way of being-interpreted (already in advance) that is typically not directly experienced as such: in it, rather than *taking* itself to be interpreting itself in a specific way, Dasein typically just understands itself as passively saddled with these indifferent public "givens." Nevertheless, its actual basis, accessible to the formalizing indication, remains a "definite manner of Dasein's *having been interpreted*," one, Heidegger says, "from out of which it lives, on the basis of which it is *lived*," in other words, a "how of its being."[66]

The crisp discussion of the "today" here compresses into a handful of pages the detailed analysis of publicity, the character of the "one" or *das Man*, and its ontological foundation in *Gerede* (idle talk) as a mode of discourse, which Heidegger would carry out in much more detail in division 2 of *Being and Time*, three years later.[67] For the present discussion, however, what is most relevant is not these specific structures themselves, but rather the way that this brief example illustrates the methodology of formal indication. As we have just seen, the ontological-temporal character of the "today" as indicated in this brief formal indication is: (4) a *formal-ontological* (and thereby temporal) character structurally existing within the total life and ontological character of Dasein; essentially *indicated* (3) on the basis of *one's own* case as it is lived (1). Furthermore, it thereby articulates specifically *how* the "today" is "given" there, in one's own case (2), namely as a definitive mode of self-interpretation that is typically inexplicit and vague.

In relation to the analysis of the "today," the specific features of the formal indication mark it off from other, more traditional forms of definition or analysis, which would not even clearly be applicable at all in this connection. Heidegger's focus is not, for example, on defining "today" as a general category, or in relating it as a *species* to a more general categorical definition of time or of the present (indeed, given the indexical nature of the term under consideration, it is not clear what could be meant by doing so). In fact, the analysis here does not elicit *categories* at all, but rather the ontologically characteristic structures that Heidegger calls, in *Being and Time*, "existentials," explicitly by way of a "sharp distinction" of them from the categories of things.[68] Nor does he attempt a definition of "today" in the sense of finding necessary and sufficient conditions, for instance (what would surely be absurd) conditions of *its being* "today" (now or at any other time). Whereas it would clearly be, for similar reasons, quixotic to attempt a general analysis of utterances involving the indexical "today" in terms of standing and language-independent truth-conditions, the evident and indicative meaningfulness of such utterances (including their capability of truth and falsehood, as uttered), points here to the possibility of, and indeed need for, the kind of formal-indicative analysis that Heidegger undertakes. This is one, in particular, suited to articulatively demonstrating the underlying character of being that is the basis for understanding and experiencing any collective or public "today," one aspect of the ontological/temporal "meaning" or "sense" of the present as such. As thus explicitly tied to, and explicative of, the demonstrative and indexical character of the phenomenon in question, the formal indication here appears capable of clarifying what general definitions and abstract analyses cannot: namely, the actual *form* of the phenomenon as presented, which is here not categorically distinct from the presentation of the (temporal) present as such.[69] The formal indication, if successful, thus points in part to the underlying formal and ontological structure of this phenomenon, the logical form of the present itself.

It is further instructive to compare formal indication with other types of philosophical definition, articulation, and reflection, and especially with some of the methods that have been advanced in twentieth-century discussions to elicit (what is called) "logical form."

For example, in that it points to underlying structures that are essentially "given" only *within and by reference to a structural-relational whole in which they are included*, the method might usefully be compared to that of *implicit definition*. In the context of axiomatic theories such as Euclidean geometry or ZF set theory, key terms (for instance, the set-theoretical "element of," "∈") are not defined *explicitly* by means of direct stipulative definition or the introduction of necessary and sufficient conditions. Rather, they are *implicitly defined* by the axioms themselves. This means that the axioms do not identify the referent of the key term with any other discrete elements or components; rather, they jointly describe its total relations to the referents of other terms within the system. What is thereby defined, then, has its *sense* only in the irreducible context of the whole system, and through the (ultimately syntactically specifiable) relations it bears to the referents of other implicitly defined terms. In this respect—that of indicating an underlying structure in terms of the irreducibly holistic context of relations that constitute it (feature 4 above)—implicit definition resembles formal indication. However, implicit definition (as it is used in particular axiomatic theories) has no evident analogue to the *first* feature of formal indication, its essential reflexivity. And its analogy to the second—the indication of how the phenomenon is *given*—is limited to the claim that the definition of implicitly defined terms is "given" along with the axioms and in the way that they themselves are. But as to how the axioms *themselves* are given, or how we should understand their potential "truth," the methodology of implicit definition by itself is silent (although, by contrast, clarifying how the holistic context in which it takes place is itself given in truth is an *essential* aspect of the formal indication). Finally, an implicit definition of a term within a particular axiomatic system could be considered to answer to the *third* feature—that of having *indicative* value with respect to what is "there"—only to the extent that the structural logic of the system in question could legitimately be considered to capture some *already existing* structure of meaning and practice. And whereas a (successful) formal indication might indeed serve to verify indicatively the legitimacy of such a formal "capture" of these preexisting structures, the implicit definition itself has nothing to say to it.

A closer, but still very imperfect, analogy might be drawn to the practice of *articulating reasons* within a broadly "inferentialist" analysis of structures of intersubjective, discursive practice.[70] Within projects of this sort, formalization in the sense of rational articulation of the (typically initially "implicit") instituted regularities of discursive practices is seen as a kind of rational (self-)reflection. This reflection is intended to bring the bases of these practices to light, as they in fact (already) are. Conceived as such an "organ of semantic self-consciousness," formalization here thus exhibits (at least some aspects of) feature 3 above: that of *making explicit* or *recognizable* what is already *there* in concrete life (here understood as the concrete reality of the particular instituted discursive practice).[71] As an essentially *holistic* type of linguistic-semantic articulation of structure, moreover, it also exhibits some parallels to feature 4 above. However, articulative explicitation in the sense of articulating reasons is less analogous to formal indication with respect to features 1 and 2. For it does not clarify the way in which the relevant practices—along with the "norms" and "statuses" they are construed as having—are *given* or constituted, except by means of a highly problematic reference to their conventional "institution." Such an institution, if conceived as an origin for the phenomena of linguistic meaning, intensionality, and the force of reasons, must apparently presuppose the being of language it seeks to explain.[72] And although it can indeed operate—at least in part—as reflection on our own practices, the methodology of pragmatist articulation does not exhibit the reflexivity that is essential to formal indication (feature 1). Nor does it thereby give a genuinely indicative *demonstration* of the structural basis of the norms and regularities constitutive of (given) discursive practices: at best, it *describes* this basis externally, by means of an account that could (in principle) be understood from a perspective quite alien to that of an actual practitioner of the practice itself. This is not so, as we have seen, for formal indication: here, what is essential to the indicative methodology is that its result be recognizable as an ontologically constitutive structure of the investigator's *own* life and being.

This methodological limitation of pragmatist articulation is further shared with other familiar approaches to the articulative definition of primarily semantic structures. These include "conceptual-role semantics," as well as, indeed, the "topic-neutral" forms of description typical of functionalist and other physicalist analyses in the philosophy of mind.[73] In envisioning a type of analysis of the phenomena that culminates in the provision of a relational and holistic structural description of them that is itself in principle accessible independently of the concrete life in which these phenomena are given, projects of this general type leave out the "interior" or first-personal aspect which is essential to the ways the phenomena are given and thus to their ways of being, themselves.[74] Although there is no need to characterize this interior aspect as inherently or ineluctably idiosyncratic or essentially private, if it goes missing from the analysis, an essential dimension of ontological clarification will inevitably be lost.

A still closer, though still not perfect, analogy might be drawn between formal indication and the methodology of *ordinary language philosophy*, as developed (for example) in the account given of it by Stanley Cavell.[75] Here, by contrast with the methodology of inferential articulation, the consideration of "what we should say when . . ." is indeed essentially *reflective* and *first-personal*. The aim of the reflection is not an indifferent description, but the clarification of what are already *for us* the meanings of our lives and practices *as* these are concretely embodied in the language we speak. As Cavell has argued, moreover, what such a methodology yields is not an exterior description but *recognition*: its stakes are not the correctness of an exterior account, but rather the acknowledgment that we practice, or withhold, with respect to others whom we thereby can treat (or not) as sharing with us a life and world. The reflective methodology of ordinary language philosophy is thus analogous to formal indication with respect to the first feature—that of reflexivity—and in its investigation of the concrete meanings of our shared life and practice, on the basis of the way these are present in the very language we speak as we ourselves have access to it, it substantially captures features 2–4 as well, at least in some of their dimensions.[76]

The limitations of the analogy here, however, start to emerge if we consider ordinary-linguistic reflection in terms of what is undoubtedly its essential and definitive resource: namely, ordinary language itself. In particular, since ordinary-linguistic reflection limits itself to consideration of *language* (in the sense of "what we should say"), it is unclear what its genuinely *ontological* status is: to what extent and how, that is, it can bring to light and indicatively clarify those structures that characterize our *being* rather than just the *language* in which it finds (typically inexplicit) expression. This is not to say, of course, that reflection on language has just *nothing* to do with hermeneutical ontology. Rather, as we shall see in more detail in subsequent chapters, the problem of the relationship of linguistic terms and categories to ontological-existential structures indicated on the basis of a hermeneutics of facticity is itself a foundational one for hermeneutical ontology. However, this problem must itself be clarified before ordinary language philosophy can be allied with formal-indicative hermeneutical definition in any conclusive way.[77]

In connection with the hermeneutic, interpretive method of inquiry that Heidegger had developed by 1923 and would continue to employ in *Being and Time*, formal indication thus provides a methodological way of clarifying, on the basis of given, factical experience, more unitary, underlying and constitutive structures of Dasein as such. But despite its basis, in each case, in the singular factical life, the formal indication remains *formal* in the particular sense of the bond that it provides back to its ontologically underlying structure and thereby to the particular phenomena characteristic of that life in their own characteristic ways of being given there. The question that Plato originally formulated as that of the basis for the figuring of forms (in the sense of ideas) in thought and sensible experience is thereby put on a new and ontologically transformed footing. Here, in particular, the question of the givenness of form is no longer that of the availability of the peculiar objects of thought that also determine the properties of extant objects, but rather that of the interpretation of the concrete phenomena of life as they are given both to experience and to thought. With this hermeneutic clarification of the givenness of the formal dimension of the constitutive phenomena of Dasein, the structure and form of their possible givenness is interrogated at the same time, so that the phenomena are neither understood in "objective" abstraction from their own proper givenness, nor made to conform or be reduced, in idealist fashion, to the conditions imposed by it. The result is a hermeneutic methodology of formalizing that can, as we shall see in more detail in the following chapters, concretely clarify the underlying phenomena of sense, truth, and time as they figure in Dasein's concrete life and are there given to be experienced and thought.

In pursuing these investigations, I shall often consider results that have emerged in the course of the twentieth century in the context of analytic philosophers' attempts to formalize logical structure, semantic meaning, and truth by means of the development of formal symbolisms, especially in relation to the analytic tradition's own reflection on the proper methods, scope, and constitutive limits of such formalization. To see these results as formally indicative—as I repeatedly shall in the chapters that follow—is admittedly to go beyond any use that Heidegger himself ever makes of formalization in the specific sense of reflection on the formal-symbolic.[78] Nevertheless, the practice has a warrant in the methodology of formal indication itself, insofar as the development of symbolism and the reflection it inherently involves about the structure and limits of linguistic expression are themselves essential *aspects* of our concretely experienced lives as the "linguistic" beings we are. To develop the implications of formalism in part through a reflection on formal symbolism, as I do in the following chapters, is not to presuppose that just every phenomenon of life can be treated in this way, or that the imposition of a formal calculus is ever, just by itself, sufficient to elicit the underlying ontological structure of the phenomenon in question. But it is nevertheless clear that the project of symbolic-formalization that various projects of "analytic" philosophy have undertaken can have an ontologically and formally indicative significance with respect to the phenomena they take up, even and indeed *especially* where these projects in fact *fail* wholly to capture these phenomena themselves.

We shall consider from this perspective, for example, Tarski's analysis of the structure of truth-definition for a particular language, Davidson's project of holistic structural interpretation of natural languages, Dummett's reflection on the formal consequences of the adoption of logical bivalence, Gödel and Turing's "limitative" results about the completeness of formal systems and the availability of decision procedures, and Wittgenstein's consideration of what it means to follow a rule. What is essential in justifying the connection of these results to the broader ontological problematic and to the broader methodology of formal indication, in each case, is just to note that "sense," "truth," "logic," "language," "meaning," "decision," "completeness," "consistency," "rule," "being," and "time" are *themselves significant terms of ordinary language.* As such, their significance can and must be understood in relation to the ultimately ontological horizon in which their use unfolds, and to which it can again be indicatively related by means of the right kind of formalizing reflection. To employ this formalizing methodology in connection with these phenomena cannot be, as Heidegger indeed already emphasizes, simply to reduce them to their description in terms of the structures or principles of a "formal" or purely symbolic logic. Nor can it be simply to subject them to the fixed structures of one or another ready-made formal-symbolic calculus. Rather, what is involved here is, as Heidegger already suggests, an essentially *reflexive* formalizing *articulation* of structures that are already "there" in advance in concrete life at the formal-ontological basis of the concrete phenomena indexed by these terms (though typically inexplicitly so) and capable of also (co)indicating their specific ways of *being* there.

4. Critique of Husserl: Time, Ideality, and the Being of the "Subject"

By 1923, Heidegger was thus already in essential possession of the distinctive methodology for the elicitation of the formal structure of Dasein that he would employ in the "Preparatory Fundamental Analysis of Dasein" that comprises division 1 of *Being and Time.* Over the next several years, his further radicalization of the methodology first suggested by Husserl for the demonstration of what were for him the atemporal categories of being, would lead Heidegger to undertake a devastating critique of his erstwhile teacher and to penetratingly reopen the most fundamental questions concealed beneath the traditional determination of the senses of being and time.

In the 1925 course "History of the Concept of Time," Heidegger focuses his explication of phenomenology on the interpretation, and deepening, of what he sees as its three most important discoveries so far. The first is the discovery of intentionality as the basic structure of "lived experiences [*Erlebnisse*] as such" that makes possible all judging, meaning, and understanding.[79] Here Heidegger emphasizes the inadequacy of representational theories of awareness and consciousness: in a simple intentional act of perception, for instance the perceiving of the chair before me, there is no secondary representation of the object but rather simply the direct presence in perception of the chair itself.[80] More generally, according to Heidegger, intentionality is deeply misinterpreted if it is understood as a relationship between the distinct spheres of the psychic and the physical.[81]

The second major discovery which Heidegger sees phenomenology as having made is that of categorial intuition (discussed in section 1 above). Heidegger here rehearses Husserl's conception of truth as *adequation*, or the "being-identical" of what is intended and what is intuited [*Identischsein von Vermeintem und Angeschautem*].[82] But already in this conception, Heidegger suggests, there is a basic and important ambiguity.[83] According to Husserl's theory, in a successful intentional act that attains truth, the meaning-intention is actually brought into coincidence with the fulfilling intuition of its object; this is the actual performance of the "act of identification" of the meant and the intuited. Is, then, truth to be identified with the (presumably atemporal) "subsistence [*Bestand*]" of the standing *relationship* of identity between what is intended *in* the act—its "content" or what is "meant" in it—and what is actually given? Or is it rather to be identified with the actual, particular, and datable *act* of bringing-together the intention and the intended? In the first case, truth will be the standing, always-possible correlate of a *possible* act of correlation or identification, but will not be identified with any one such act. In the second, by contrast, it will be an aspect of the *achievement* of such a temporally specific act itself.

In fact, according to Heidegger, these two possibilities for understanding truth have never been clearly distinguished within phenomenology. But the confusion can be avoided only if we understand truth in a third, more basic sense, one that refers neither to a subsisting relation of abstract identity between intention and fulfillment, nor to the actual performance of any act of identification. This third and (according to Heidegger) more basic sense of truth is one *grounded in the intuited entity itself,* which itself "provides the demonstration" of the truth of any assertion or statement about it.[84] This is, Heidegger suggests, the concept of truth as *aletheia,* or as the structure underlying the unconcealment of the entity itself.[85]

Further interrogation of this structure now leads Heidegger to what is perhaps his most decisive and central criticism of Husserl's phenomenology as a whole. Husserl's development of categorial intuition, in connection with the phenomenological theory of intentionality and the later distinction between the real (noetic) and the ideal (noematic) aspects of intentional acts, has elicited an original givenness of what Husserl understands as the "ideal." This is the actuality of forms and categories that are given to consciousness in categorial intuition without being (for Husserl) in any way real or temporal. But what Husserl has failed to do is to clarify fundamentally the relationship of this posited "ideal" realm to the *real* temporal flow of consciousness itself.

Accordingly, Heidegger ventures to raise the question of the ideal and the real in phenomenology in a sense more radical, and penetrating, than Husserl has been able to. In the critique of psychologism begun in the "Prolegomena to Pure Logic" at the beginning of the *Logical Investigations,* Heidegger suggests, Husserl has indeed decisively pointed out the fundamental shortcomings of a naturalism that remains rooted in "blindness to the nonempirical" and to "propositional content [*Satzgehalt*] as such," which means, for Husserl, "*ideal* being."[86] In criticizing psychologism in these terms, Heidegger suggests, Husserl has pointed to the "fundamental error" involved in psychologism's failure to recognize the "difference [*Unterschieds*]" of a "basic differentiation in the being of entities [*Grundverschiedenheit im Sein des Seienden*]" which is indeed the proper ontological basis of logic.[87] But although the critique of psychologism must certainly be deemed successful in pointing out fundamental distinctions of being wholly overlooked by naturalistic accounts, Husserl's own understanding of the sense and structure of judgments as founded in "ideal" being is insufficient. In fact, Heidegger suggests, the phenomenological critique of psychologism, in the form in which Husserl has pursued it, has in certain ways even *increased* the "danger" of a fundamental misinterpretation of the relationship between "mental" acts of judgment and what figures there as their "ideal" content.[88] This becomes clear when the question of the *temporality* of the "ideal" is itself made explicit. In particular, despite the fundamental usefulness of the methods and tools of phenomenological research that he has introduced,

Husserl's formulation of the distinction between act and content has simply presupposed the "Platonic" distinction between the ideal realm of form as *timeless* and the *temporal* realm of the sensory. The result is that he has ultimately failed to clarify the obscure sense or senses of being that actually underlie the kinds of demonstration that phenomenology can achieve.

Heidegger further suggests that Husserl's simple presupposition of the ideal/real distinction, which he fails along with the rest of the philosophical tradition to clarify, is in fact grounded in another failing, one which is ultimately responsible for what Heidegger sees as the decline of phenomenology into transcendental idealism, subjectivity, and an ultimately "personalistic" attitude that fails fundamentally to elucidate the very categories it presupposes. In particular, Heidegger suggests in the 1925 course, Husserl finally understands "pure consciousness" as the *ultimate* region of being, a region in which all others—the being of the real and spatiotemporal *as well as* that of the ideal—are constituted and given sense. But in so doing, Husserl has in fact *failed* to inquire into the "being of this region," the actual ontological status of the "ur-region" of consciousness itself.[89] Specifically, in thinking of consciousness as the ultimate region of the givenness of and constitution of being, Husserl has failed to inquire into the actual possibility of the subject of such consciousness himself to exist as a "real human being," concretely and factically there in a world.[90] Within this framework, moreover, the reality of the *psuche* and its concrete acts of intentionality, though of course rigorously distinguished from the ideality of their contents, nevertheless themselves appear to have the kind of status and temporality of natural objects and processes, wherein "every entity is taken *a priori* as a lawfully regulated flow of occurrences in the spatio-temporal exteriority of the world."[91] What is missing in this whole conception is, once again, a more penetrating inquiry into the kind of being of the entity thus specified, and the "primary experience of the being of man [*primäre Erfahrung des Seins des Menschen*]" characteristic of it.[92]

The redirection of the inquiry toward the actual being of Dasein has the further consequence that in it, truth can no longer be treated as resting in the *agreement, correlation,* or *correspondence* of entities. Whether the relationship in which truth is thought to consist is seen as that between a physical object and its mental or psychological correlate, or rather (as with Husserl) as that between an *ideal* content and its fulfillment in the *real* psychic act, the whole conception of truth as grounded in relation here must cede, Heidegger argues, to a more basic understanding of truth in terms of the unconcealment of entities themselves, as well as the broader structure of "world" that makes this unconcealment possible.

In the "Logic" course, Heidegger further clarifies the implications of such an understanding of truth for the availability of entities and situations and for the structure of the logic of assertions and propositions. In particular, the truth and falsity of all propositions and assertions is based in a *prior* level of the availability of beings. This availability is itself grounded in the "world-openness" of Dasein which also, reciprocally, makes "the world" open for it.[93] This allows Heidegger to oppose any correspondence theory of truth, and indeed suggest that there is no need to conceive of truth as a "relationship" between entities at all:

> Truth is not a present relationship [*vorhandenes Verhältnis*] between two entities that are present [*zwischen zwei Seienden, die vorhanden sind*], for instance as psychical and physical. It is also no coordination [*Zuordnung*], as one loves to say these days. If it is a relationship [*Verhältnis*] at all, it is one that has no analogies with any kind of relation [*Beziehung*] between entities. It is—if one can put it this way—the relationship [*Verhältnis*] of Dasein as Dasein to its very world, the world-openness [*Weltoffenheit*] of Dasein, that is itself uncovered—Dasein whose being toward the world [*Sein zur Welt*] itself is disclosed [*aufgeschlossen*] in and with this being toward its world [*Sein zu ihr*].[94]

As we shall see in more detail in chapter 3, the antirepresentationalist conception of the basis of truth that Heidegger here propounds points also toward an ontological clarification of the basis of the possible truth of assertoric *sentences* capable of avoiding the problems to which correspondence and representational accounts of sentential truth have characteristically been prone. These are, primarily, the problems of avoiding, on the one hand, an infinite regress of subjective representational contents (such that every content, as represented, must itself apparently be "interpreted" by another represented content, and so forth), and on the other, a "Myth" of the *conceptually unstructured* "Given."[95]

This clarification of the original possibility for worldly existence and the disclosure of significance is just what Heidegger takes himself to have gained through his description of Dasein and the hermeneutic method of displaying it by means of formal indication of its structures. Nevertheless, as Heidegger recognizes, all of this still leaves open the essential question of *temporality*. In particular, Husserl along with the entirety of the tradition conceives of the relationship of the ideal to the real as the relationship between two regions determinable most basically in terms of their temporality: the ideal is the region of the timeless, whereas the real is in time. The problem at the root of Heidegger's objections to Husserl's project is that of finding a way to cross the gap between these two "regions." This is just the problem of finding a way for what is conceived as atemporal, ideal form really to enter into, to be given in, temporally flowing reality. In fact, Heidegger suggests, once we conceive of the givenness of form as a matter of the figuring of atemporal ideals *within* temporal reality, this problem is basically insoluble: there is no coherent way to bridge the gap between the ideal realm of form and the real realm of temporal life it shapes. But this does not at all mean that Heidegger wishes simply to absorb form into the temporality of empirical life or return to the psychologistic assimilation of the givenness of form to datable acts of the empirical psyche. Rather, he undertakes instead to interrogate in a more basic way the very givenness of time itself.

To clarify what is involved in this interrogation, Heidegger once again draws centrally upon Husserl's phenomenological methods, developing them in the direction of a deeper posing of the question of the very sense and meaning of being as determined by time. In the "History of the Concept of Time" course, Heidegger presents as a third fundamental breakthrough of phenomenology the discovery of the "original sense of the *a priori*." Assuredly, the notion of the *a priori* as "that which already always is the earlier [*was von früher her schon ist*]" is already marked in Plato's understanding of the distinctively prior mode of existence of the forms, and in Descartes and Kant this *a priori* is thought in terms of the priority of the subject, as that which comes before and forms the basis for knowledge of objects.[96] Nevertheless, according to Heidegger, it was left to phenomenology to identify, with its distinctive invocation of categorial intuition and its essential reference to the demonstration of ideality, an *a priori* that is "not limited to" and indeed "has primarily nothing at all to do with subjectivity."[97] In fact, Heidegger suggests, despite Husserl's own official understanding of the categories as constituted by "transcendental subjectivity," the real significance of the discovery of categorial intuition is ultimately to make it clear that "something like the highlighting of ideas occurs both in the field of the ideal, hence of the categories, and in the field of the real."[98] In this sense, the *a priori* is neither, Heidegger suggests, the determinant of a specific mode or region of entities, nor of a particular kind of comportment [*Verhaltens*] or capacity of subjects. Rather, it is an exemplary indication of one of the senses of being as such, a "*title for being* [*ein Titel des Seins*]."[99] Here, the essentially *temporal* character of the "priority" of the *a priori*, as "earlier" and "already," is itself essential in articulating the real ontological significance of an invocation of the realm of the ideal or of the ideality of contents conceived, within the scope of their metaphysical determination, as timeless in themselves.

Heidegger thus sees in Husserl's discovery of the "original sense of the apriori" a first and leading indication of the radical possibility of something like a disclosure of the *temporal* sense of being as such, one that owes nothing to the being of any specific entity, including that of the subject.[100] In particular, if, as Heidegger suggests, the historical "discovery of the a priori is really connected or actually identical with the discovery of the concept of being [*der Entdeckung des Seinsbegriffes*] in Parmenides or in Plato," it is necessary in reconsidering the "prevalence of this particular concept of being" to reopen a radical interrogation into the temporality presupposed in it, the concept of being as presence that has constantly been presupposed since the Greeks.[101] Heidegger is thus led, finally, to take up once more the question of the possible givenness of the *a priori* dimension of the formal determination of the structures of the disclosedness and meaning of being, now along the guideline of the essential question of the givenness of time.

3

The Ontology of Sense and "Transcendental" Truth: Heidegger and Davidson

Over the past two chapters, we have considered the problem of the relationship of *truth* and *time* in a general form (chapter 1) and articulated a methodology (chapter 2) whereby this problem can be *ontologically* clarified: brought out and articulated, that is, in its relationship to being and to the ontological difference between it and beings. This is the methodology of *formal indication* and, more broadly, indicative hermeneutic *reflection* on the consequences of formalism. As we have seen, to employ this methodology in connection with the question of truth and time is, at least in part, to consider the implications for it of the attempt to theorize these phenomena both ontologically and formally, including, decisively, the inherent *limits* of both kinds of positive theorization. In this chapter, I consider specifically the contemporary prospects for an understanding of *truth* that draws both on the outcomes of Heidegger's questioning of being and on twentieth-century "analytic" inquiry into language and its structure. As I argue here, in order to be successful, such an understanding must accommodate, on the one hand, Heidegger's conception of linguistic truth as grounded in an ontologically prior phenomenon of unconcealment or disclosure, and on the other, the patterned structure shown in Tarski's schema for the structure of truth-predicates for particular languages. Because the fullest development of the implications of this structure in relation to "natural" languages is Donald Davidson's, I here explore the prospects for reconciling Davidson's conception of truth with Heidegger's within a unified methodological framework of hermeneutic interpretation and phenomenological demonstration.

At first glance, the two projects in which these specific conceptions arise can appear to be methodologically quite at odds with one another. Whereas Heidegger, eschewing formal symbolism and calculi, develops the implications of an *ontologically* prior phenomenon of the unconcealment of entities and the disclosure of world, Davidson's interpretive project applies the Tarskian framework first used to model truth-predicates for formal languages, thus privileging the *linguistic* truth of sentences as basic. The difference in approaches to truth exemplifies a familiar and more general tension between characteristic methods of analytic philosophy and those of phenomenological ontology. Whereas the analytic philosopher is likely to look to the *clarification* of the structure of logic or language as the basis for any possible illumination of the concepts of truth and meaning, the phenomenologist characteristically seeks a concrete *demonstration* of the "matters themselves" underlying the concrete phenomena as they are actually *given*. Following the first approach, many analytic philosophers have taken the kind of truth exhibited by *assertoric sentences*, *propositions*, or other linguistically shaped items as basic, whereas the second approach has led phenomenologists to point toward a "prelinguistic" or "nonlinguistic" phenomenological basis for this kind of truth in the actual *appearing* of things. The two approaches find prominent examples in the projects of Davidson and Heidegger, leading the latter to his long-standing critique of the "logical" assumption of the primacy of the "assertion" or assertoric sentence in the analysis of truth, while leading the former to propose an account of linguistic truth that may seem essentially committed to just this assumption. Despite these apparent differences, however, I argue here that the approaches can indeed be reconciled in such a way as concretely to indicate a unified *problematic* of truth at the basis of both conceptions.

In the sense developed here, the formal indication of the problematic, in which the central contours of a phenomenon first become evident, is neither a *definition* of the phenomenon nor a general *theory* of it. It is, rather, a clarifying demonstration of the phenomenon of truth as it presents itself, grounded in an interrogative questioning of it, and thereby pointing to the determinate points of its possible conceptual articulation, including importantly the demonstration of the inherent points of aporia or theoretical blockage that may ultimately render untenable the hope for a single and adequate theory. Heidegger and Davidson both effectively argue that there is no direct route from empirical facts, ontic configurations of entities, or the epistemic or cognitive capacities of agents, subjects, or communities, to the structure of truth as such. If this is right, then there are good reasons for thinking that truth can neither be defined in a unitary way nor described by a single, complete, and consistent theory applicable to all languages and situations. Nevertheless, as I shall argue, it remains possible that the underlying phenomenon can be indicated both in its concreteness and in its givenness by means of a twofold formal demonstration. The demonstration would show its underlying ontological and logical-semantic structure in relation to its manifold appearances in languages and concrete lives.

1. Davidson and Heidegger on "Transcendental" Truth

There are criteria for the intelligibility of *things* and *events*, relative to empirical situations or domains of inquiry; and there are formal/structural *truth-predicates* employed in natural languages, relative to those languages, and constitutively linked to their own structures of linguistic sense. But to ask after the possibility of a *unified* semantic/ontological structure of truth is to wonder whether there is, behind each of these and uniformly conditioning both, a *phenomenon of truth as such*. Such a phenomenon would underlie the articulation of criteria of the genuineness of entities and phenomena in particular ontic domains, *as well as* the conditions of the truth of sentences in particular languages. In this chapter, I designate as *transcendental* any conception of truth that is, in this way, not limited with respect to either languages or domains of entities: that is, any conception of truth that aims to indicate its nature or clarify its structure *prior to* its bearing within a specific and well-defined field of inquiry, or a particular natural language. Beyond this, nothing much is meant to turn on the terminology.[1] The point is just to indicate the universality of what might otherwise be called a *general* concept of truth, correlative to a single phenomenon of *truth in general* or as such, were it not problematic (as, we shall see, it is) to understand the relationship of truth to its determined situational instances as that of any kind of genus (even a "maximally general" or "preeminent" one) to its more narrowly determined species.

As we shall see, both Davidson and Heidegger see truth as grounded in the more basic underlying hermeneutic *situation* in which we holistically "make sense" of languages and the world, and in the transcendental phenomenon of truth that is operative there. Here, indeed, for both philosophers what basically qualifies truth for a "transcendental" role with respect to particular languages or situations is the unitary way in which it guides interpretation, whether it is the interpretation of the structure of language or of facticity itself that is at issue. What grounds are there for thinking that Heidegger's and Davidson's specific conceptions of truth, despite their obviously different motivations, can indeed be brought together into such a single, hermeneutically oriented conception? To begin with, it is helpful to note that there are at least three general *negative* features of both philosophers' accounts of truth on which they agree, in contrast with many other contemporary theories and accounts. First, both philosophers reject *correspondence* theories of truth.[2] Second, both philosophers reject *coherence, antirealist, and other epistemically based* theories of truth.[3] Third, both philosophers reject the existence of propositions, Fregean thoughts, ideal contents, or other *timeless* entities as the primary truth-bearers.[4] Beyond these connections, however, both philosophers, despite the differences in how their accounts are situated, point positively to the need for an account of the phenomenon as a "general" or "transcendental" one in the sense described.

In his last, posthumously published book, *Truth and Predication*, Donald Davidson suggests that the use of Tarskian truth-definitions for particular languages to produce theories of meaning for them, in the context of radical interpretation, depends on a preexisting understanding of a *prior* concept of truth, not itself specific to any particular language. Familiarly, on Davidson's long-standing conception, a "theory of meaning" for a natural language recursively embodies a compositional structure of assignments of meaning to the language's primitive predicates and singular terms.[5] This structure of meaning can be embodied by a theory that yields as consequences all of the true T-sentences for a particular language. This is the structure described by Tarski in "The Concept of Truth in Formal Languages" as the one that must be exhibited by any extensionally adequate definition of the term "true" as it is used in a (formal or natural) language.[6] According to Tarski, any such definition will be adequate only if it implies all sentences of a certain form, what he calls form (T):

X is true, if and only if, p.

Here, "p" stands for any sentence of the language, and "X" is to be replaced with a name or designative description of that very sentence, formed by enclosing the sentence itself within quotation marks, or by some other naming device.

Thus, for instance, a Tarskian theory of truth for English will imply that

"Snow is white" is true (in English) if and only if snow is white.

Tarski suggests, in "On the Concept of Truth in Formalized Languages," that a definition of the truth-predicate that implies all the instances of the T-schema will be both "materially adequate" and "formally correct": that is, it will capture the actual behavior of the truth-predicate for the language and will do so in a way that reveals its underlying formal structure.[7] And as he goes on to show, such a definition can in fact be constructed from that of a more primitive "semantic" relationship, the relationship of "satisfaction."

Formally speaking, the relation of satisfaction can be understood as coordinating primitive singular terms of a formal language to individual objects, and primitive predicates to sets and sequences of objects. More intuitively and loosely, the idea underlying it is essentially (as Tarski put it in the 1944 article "The Semantic Conception of Truth") that of a maximally general *semantic* relationship, or relationship between linguistic signs and objects in the world. In particular, satisfaction is a *generalization* of the "semantic" relationship that (on Tarski's account) holds between a true sentence and the world, which applies (in its more general form) not only to sentences but also to propositional functions with open variable places. For example, the propositional form "*x* is an apple" is not a sentence, because of the appearance within it of the unbound variable *x*, but we can nevertheless consider it to be *satisfied* by a specific range of objects, namely all those (and only those) that are—in fact—apples. Similarly, the propositional form "*x* is taller than *y*" is understood as satisfied by all (ordered) *pairs* of objects wherein the first is actually taller than the second. Given a development of the role of logical quantifiers and truth functions in the syntax of the language, this yields the result that quantified sentences (such as "there exists an apple" or "everything is an apple") are, if true, satisfied by *all* sequences of objects, and if false, by none. And given the specification of the satisfaction relations, the definition of the truth-predicate can be built up recursively from them. In this way it is possible actually to define the truth-predicate (which must characterize an infinite number of possible sentences) from a finite set of axioms (the specification of the satisfaction relations for the [finitely many] basic terms of the language).

In Davidson's project of analyzing the semantical structure of natural languages, the order of explanation that characterizes Tarski's truth-definitions is, in a certain way, reversed. Rather than beginning with primitively specified satisfaction relations for specific formal languages in order to build up the recursive structure of truth for the language, Davidson (following Quine) envisages the radical interpreter beginning with the project of interpreting an already existing natural language at first completely obscure to her, and working to reconstruct its underlying structure from the attitudes of holding-true and rejection of sentences exhibited by its speakers. Nevertheless, the recursive structure underlying the true T-sentences remains the primary object of investigation, and Davidson argues that (as for Tarski) this compositional structure must be adequate in the sense that it yields as deductive consequences the whole *corpus* of T-sentences for the language in question. A speaker's understanding of the language may then be considered equivalent to her knowledge of this recursive structure, and an interpretation of it in another language may be considered a precondition for successful understanding of speakers of the first language by speakers of the second.[8]

It is here that Davidson invokes, in *Truth and Predication,* the need for a *general* concept of truth in the practice of interpretation itself. For the interpretation of a specific language, if it is to recover (what is intelligible as) the structure of a truth-definition for that language at all, must presuppose and be guided by such a more general concept. This general concept of truth is thus to be distinguished from the particular Tarskian definitions of the truth-predicates for individual languages that Davidson conceives as underlying specific "theories of meaning" for those languages. For these definitions themselves do not indicate what the various truth-predicates have in common; but it must be possible to see them as having a deeper, common structure if we are to use them in the linguistic interpretation of beliefs and meanings at all.[9] Davidson puts the matter this way:

> My own view is that Tarski has told us much of what we want to know about the concept of truth, and that there must be more. There must be more because there is no indication in Tarski's formal work of what it is that his various truth predicates have in common, and this must be part of the content of the concept. It is not enough to point to Convention-T as that indication, for it does not speak to the question of how we know that a theory of truth for a language is correct. The concept of truth has essential connections with the concepts of belief and meaning, but these connections are untouched by Tarski's work.[10]

In particular, while Tarskian theories for specific languages indeed point to a general structure which must be fulfilled by any systematic account of meaning, showing the "kind of pattern truth must make, whether in language or thought," neither the particular Tarskian truth-definitions nor this general pattern suffices to exhaust the underlying concept of truth as it must in fact be presupposed in the practice of actual linguistic interpretation.[11] This understanding is supplied, in each case, by the systematic interpretation of the speakers' utterances in radical interpretation, but it is dependent in that context on the prior grasp of a non-language-specific concept of truth as such, which we must be able to have without yet having any detailed explicit understanding of the structure of the language. It is this "untutored grasp" of an underlying concept of truth which we then, according to Davidson, draw on in interpreting a language by attributing truth-conditions to the utterances of its speakers, and which further convinces us that the structures of Tarski's formal machinery, in application to particular languages, "pretty much accord" with the antecedent concept.[12]

As Davidson further argues in *Truth and Predication*, the necessity of such an effectively presupposed concept of transcendental truth for interpretation bears against the claims of those who have seen in Tarski's conception the warrant for *deflationary* or *redundancy* accounts of truth. On these accounts, there is nothing much more to say about truth than to point out that truth-predicates function disquotationally, or that to say that something is true is just to assert it.[13] But if a concept of transcendental truth is indeed prerequisite for interpretation, the Tarskian truth-predicates may be seen as having, through their presupposition of it, "further essential properties" beyond just those actually directly involved in the Tarskian theories for specific languages, even if these theories themselves make no use of these further properties.[14] Nevertheless, Davidson argues that, even if this is so, it would be futile to attempt to *define* truth in this general sense, and all historical attempts to do so have in fact accordingly failed. For since truth in the sense in which it is necessarily presupposed in successful interpretation is one of the simplest and most basic semantic concepts we possess, it would be quixotic to attempt to *define* it in terms of supposedly more basic or foundational ones (including, Davidson suggests, concepts such as those of reference, "correspondence," "coherence," or any other such specialized philosophical notion). What we *can* do, however, is to make the underlying and non-language-specific concept of truth *clearer* by considering its essential relationships with other basic semantic concepts as well as with sentential predication itself. This is just the kind of reflective inquiry that takes place in Davidson's own analysis of the implications of radical interpretation, and he sees it as at least implicit in Tarski's own thinking about what the latter called the "semantical" conception of truth. What is to be clarified in the inquiry is, then, just that hermeneutic concept of truth which, as Davidson argues, must be able to be presupposed in practice in any interpretation of another's utterances, insofar as interpreting them involves ascribing truth-conditions at all, but which can also be formally captured in the pattern of T-sentences which systematically connect these conditions to structured utterances in a particular case.

It is also clear that Heidegger's treatment of truth as "unconcealment [*Unverborgenheit*]" considers the underlying phenomenon of truth as "transcendental" in the sense I have described. In *Being and Time*, this account largely takes the form of a description of the "original" phenomenon of truth as uncoveredness [*Entdecktheit*].[15] Thus understood, truth as unconcealment or uncoveredness is "prior" in several senses, but perhaps the most central of these is its presuppositional character with respect to the specific phenomenon of a (linguistic) assertion [*Aussage*]. Thus, in section 44 of *Being and Time* (the section that concludes division I's "Preparatory Fundamental Analysis of Dasein"), Heidegger explains the truth of *assertions* as grounded in their uncovering or unconcealment of entities:

> To say that an assertion "*is true*" signifies that it uncovers the entity as it is in itself [*an ihm Selbst*]. Such an assertion asserts, points out, "lets" the entity "be seen" (*apophansis*) in its uncoveredness [*Entdecktheit*]. The *Being-true* (*truth*) of the assertion must be understood as *Being-uncovering* [*Entdeckend-sein*] . . . "Being-true" ("truth") means Being-uncovering [*Wahrsein (Wahrheit) besagt entdeckend-sein*].[16]

This uncovering is itself grounded, according to Heidegger, in the structure of Dasein as well as that of the world. In particular:

> Uncovering is a way of Being for Being-in-the-world . . . What is primarily "true"—that is, uncovering—is Dasein . . .
>
> Our earlier analysis of the worldhood of the world and of entities within-the-world has shown . . . that the uncoveredness [*Entdecktheit*] of entities within-the-world is *grounded* in the world's disclosedness [*Erschlossenheit*]. But disclosedness is that basic character [*Grundart*] of Dasein according to which it *is* its "there."[17]

Heidegger is here thus concerned with a "most primordial" phenomenon of truth as uncoveredness that has two holistic aspects. First, the most primordial phenomenon of truth is grounded in (or even identifiable with) disclosedness as the "basic character of Dasein." Second, this basic character—the disclosedness of Dasein—is also identifiable with the disclosedness of the world, what underlies the possibility of any showing or appearing of entities *within* the world.

According to Heidegger, more generally, the possibility of predicative assertion in language has its condition of possibility in a more basic phenomenon of interpretive disclosure.[18] The most basic underlying structure of *linguistic* assertion is characterized as that of an "apophantical as" in which an assertion describes or characterizes an entity *as* something or *as* being some way: for instance, as having some particular feature or characteristic, or standing in some relationship to another entity. But this "apophantical as" of the linguistic assertion is itself, according to Heidegger, ontologically founded on a more basic "as"-structure of hermeneutical *understanding* or *interpretation* [*Auslegung*].[19] This more basic " 'as'- structure," whereby any entity is disclosed *as* something or other, always characterizes, in a fundamental way, any possible understanding or interpretation of entities. This is the case, in particular, already when entities are disclosed in "concernful circumspection [*besorgenden Umsicht*]" as ready to hand [*zuhanden*], prior to any explicit thought or linguistic assertion about them. In such circumspection, for example in handling a hammer, there need not be any explicit judgment or linguistic assertion, but it is nevertheless possible for an entity to be interpretively disclosed *as having* a particular character (for instance, the hammer as "too heavy").[20] Nevertheless, the original "as"-structure of hermeneutic understanding can under certain conditions become transformed into the *explicit formation* of an assertion. For instance, by way of a transformation in our way of being "given" the object, the "fore-having" which already characterized the hermeneutical disclosure of the hammer as hammer is changed over into the "having" of something present at hand, which can now be the "about which" of an explicit assertion.[21] The basic "existential-hermeneutical 'as'" of circumspective interpretation is thus modified into the "apophantical" "as," which makes it possible to formulate any explicit assertion about the object.

The more basic existential-hermeneutic "as"-structure, as it operates in "everyday circumspective interpretation" (with or without an explicit, thematic focus), itself breaks up into three "fore"-structures that jointly connect the individual entity to the *total* context of its involvements.[22] First, there is a "fore-having [*Vorhabe*]" whereby the totality of involvements is always already (in some sense) "understood." Second, there is a "fore-sight [*Vorsicht*]," which begins to separate from this total context of involvements the specific entity in question and makes it capable of being conceptualized. Finally, there is a "fore-conception [*Vorgriff*]," which "decide(s) for a specific way of conceiving" the entity and thus "can be drawn from the entity itself, or . . . can force the entity into concepts to which it is opposed [*widersetzt*] in its manner of Being."[23] The threefold fore-structure of understanding is itself "existentially-ontologically" connected to the basic phenomenon of projection [*Entwerfen*], whereby entities are "disclosed in their possibility [*ist Seiendes in seiner Möglichkeit erschlossen*]" by Dasein.[24] This involves that entities are "projected upon the world [*auf Welt hin entworfen*]"; "that is, upon a whole of significance [*ein Ganzes von Bedeutsamkeit*], to whose reference relations [*in deren Verweisungsbezügen*] concern, as Being-in-the-World, has been tied up in advance."[25] It is the projective relation of Dasein to this totality of significance that allows entities to be understood with respect to their distinctive ways of being.

Heidegger's understanding of the most basic common precondition of assertoric and nonassertoric truth thus involves a general phenomenon, that of the "existential-hermeneutic as," which is further characterized *both* as the foundation of the possibility of interpretation of the being of entities in their specific domains of projective appearing *and* as the ultimate underlying basis of the structure of linguistic predication. As grounding the possibility of the truth of assertions and entities but not itself limited to particular languages or ontic domains, Heidegger's conception of truth thus is, like Davidson's, a "transcendental" one. Moreover it is, like Davidson's, hermeneutic. It is grounded in an underlying conception of interpretation that links truth systematically to the intelligibility of entities and sentential truth-conditions as well as to the *holistic* intelligibility of their shared maximal context, the world as such.

For both philosophers, in fact, the idea of the hermeneutic basis of transcendental truth in the sense I have discussed points to a threefold holistic connection among truth, entities, and predication which must be decisive in indicating the structural contours of any successful conception of it. For to speak truly about an entity is to predicate something truly *of* it; and what is truly predicable of an entity characterizes what it *is* (in the "predicative" sense of "is"). Accordingly, the idea common to both philosophers of a transcendental concept of truth hermeneutically presupposed in actual interpretative practice points, in both cases, to the determinate connection of that transcendental concept to the underlying structure of predication in sentences, on one hand, and to (what is called in the Heideggerian jargon) the "being of beings," on the other. Such an underlying phenomenon of truth, indifferently "linguistic" and "nonlinguistic," is not to be construed as the general type or overarching genus of which particular criteria of truth or linguistic theories of meaning are more determined species or instances. It is, however, the hermeneutic and problematic basis on which the essential connection between the structures of linguistic predication and the determination of the being of entities can be clarified by means of an interrogative inquiry into the more original "connections" of truth, sense, and being.

2. Hermeneutics of Truth: A Twofold Picture

Beyond these initial similarities, more positive grounds for rapprochement between Davidson's and Heidegger's conceptions emerge on closer consideration of *how* the underlying hermeneutic structure of truth is linked, for both, to the specific structure of *linguistic predication*. For Davidson, as we have seen, Tarski's structural approach to truth plays an essential role in the project of giving a theory of meaning for a natural language. Beyond this, however, Davidson suggests that we can find in it essential ingredients for an actual solution to the ancient problem of the predicative *unity* of a sentence, or the problem of how the separately meaningful elements of a predicative sentence come together to produce something unified and evaluable as true or false. In the second part of *Truth and Predication*, Davidson considers attempts by Plato, Aristotle,

Hume, Leibniz, and Russell to explain this truth-evaluable unity in terms of the combination or synthesis of such separately meaningful elements.[26] Each of these historical attempts fails, he argues, either by failing to explain the actual basis of the unity of the sentence or by doing so in a way that does not account for how this unity yields something evaluable as true or false.

On Davidson's telling, though, Tarski's conception of truth provides the essential structure that is needed to account for the unity of the proposition without inviting the problems of infinite regress and explanatory idleness that seem recurrently fatal for the earlier theories. In particular, Tarski is able to succeed where others have failed by providing a systematic way, through his recursive truth-definitions, of linking the meanings of the smallest meaningful parts of sentences (the individually referring terms and predicative expressions) systematically with the truth-conditions of those sentences as whole. The key concept underlying this possibility of solution, Davidson argues, is Tarski's concept of satisfaction.[27] It is this concept that allows the theorist to characterize the circumstances in which entities or sequences of entities are assigned to the variables in an open sentence which would make the sentence true if the variables were replaced by names for those entities and sequences. In this way, the systematic truth-conditional structure of the language becomes accessible to theoretical reconstruction in the form of a Tarskian truth-theory. As Davidson suggests, indeed, there is no other structure that could systematically elucidate the essential structural relationship of predication with truth, in such a way as to account for the infinite possibility of forming truth-evaluable predicative sentences that every natural language affords.

But this does not mean, as Davidson emphasizes, that a privileged relation of satisfaction holding between singular terms and particular objects is presupposed. Rather, the application of the Tarskian pattern to natural languages aims to discern how the systematic pattern of truth-evaluable sentences itself demands that singular and other terms have *semantic roles* requiring them to be satisfied by particular objects. In making this application, the theorist does not presuppose a specific concept of satisfaction, or any other referential or reference-like relationship, but rather only the general (or "transcendental") concept of truth.

As we have seen, Davidson thinks it is idle to hope for a definition of this concept, in terms of either satisfaction or any other relation or property. Nevertheless, we can *use* the general concept to illuminate the structure of predication. The result is a general method that allows us to characterize, for any predicate in a language, the conditions *under which* it is true of any number of entities, and thereby to use the predicative terms of the language to quantify over "endless" unnamed entities as well.[28] As Davidson in fact points out, however, this leaves the "nature" of the satisfaction "relation" itself unclear. It is not clear, for example, from what independent perspective *it* could be specified, or how it might be possible, independently of the pattern of truth-values exhibited by the sentences of a language as a whole, to "fix" the satisfaction relations themselves. Indeed, as Davidson argues, in the context of the interpretation of a natural language, it is actually impossible to "fix" these relations in advance of developing an interpretation of the language as a whole (for the interpreter does not have independent access to "primitive relations" between words and objects, or general terms and their extensions). This is why Davidson suggests that in working out an actual interpretation, the satisfaction relations must be seen as an outcome of the systematic pattern of truths, rather than as a substantive basis for them. In adopting this practice, rather than "building up" the concept of truth, as Tarski does, for a particular language from specified satisfaction-relations, we are thus using our *general* concept of truth (the "most basic semantic concept we possess") to provide an explicit reconstruction of the structural basis of meaning for the particular language at hand, *including* the particular "satisfaction" or other reference-like relationships we interpret it as involving.

Without disputing the actual hermeneutic utility (and even necessity) of this general sort of picture of what is involved in interpretation (on which truth is primary and reference-like relations such as satisfaction are secondary), one must note that it is actually in substantial initial tension with Davidson's solution to the problem of predication itself. This solution turns (as we have seen) on Tarski's use of the general idea of satisfaction as the *basic* explanatory notion in characterizing the structure of a language overall. In such a characterization, the truth of a predicative sentence is treated as consisting in the predicates being *true of* the entities named by its constants or "quantified over" by its variables. Here, though, "true of" is simply a formulation of the general relation of satisfaction as it applies to predicates in relation to entities or sequences of entities, and Davidson's idea is that its specification in terms of this relationship is alone capable of avoiding the problems to which historical theories of predication are recurrently prone. But for all Davidson says, it remains in a certain sense obscure what is meant by a predicate's being "true of" an entity. In particular, because of the way that this is, for Davidson, conceptually dependent on the *primitive* and *unanalyzed* notion of truth (rather than conversely), we have no specific characterization (nor, Davidson suggests, could we have one) of the *actual* relationship to which the formulation points, and which is clearly at the basis (in one sense) of the possibility of predication itself within the framework Davidson inherits from Tarski. Even as central a matter as what satisfaction is a relationship *of* (whether, for instance, of words to words, or words to things, or things to properties) is not made any clearer simply by referring to the relationship, deep-seated though it is, between truth and satisfaction that figures in the form of Tarski's theories.

A way of putting the issue, using terminology that is not Davidson's, is that despite its obvious structural merits, the suggested solution leaves the basic relationship, and hence the actual structure of predication itself, *phenomenologically* or *ontologically* unclear. We have from Tarski the basic idea that we illuminate the structure of predication by saying that a sentence is true if its predicates hold of, or are true of, its objects. But what can we say, guided by the *general* concept of truth, about *what it is* for this to be the case? My suggestion is not, here, that a specific relation or kind of relation between words and objects would have to be specified or determined, quite independently of a specific or general idea of truth, in order for Davidson's solution to the problem of predication to be successful. Indeed, Davidson has given us good reason to think that there is no such relation to be found. Nor is the point that, in order for any specific Tarskian truth-theory for a particular language to be verified as holding of that language (as it is actually spoken), we would somehow need access to the word-object relations of that particular language (access which, as Davidson has successfully argued, we cannot generally have) independently of our interpretation of its distinctive pattern of truths.

The suggestion is, rather, that, even agreeing with Davidson that we have no access to the "satisfaction relation," in particular cases or in general, independently of our grasp of the pattern of truths, something like a transcendental idea of this relation nevertheless plays an essential role in producing the *general* form of solution to the problem of predication that Davidson draws from Tarski. And this idea of something's being *true of* something, to which Davidson's solution essentially appeals, needs further illumination in terms of this *transcendental* concept of truth if the solution is going to shed any significant light on the *nature* of predication (rather than just act as a structural placeholder for such an illumination). For it is, on Davidson's account, *this* concept which must ultimately be appealed to if we are to illuminate the nature of predication itself in general. And it is clearly this concept to which we must relate predication if we are to illuminate (what he treats as) the *general* structure of predication itself, across new cases and changes in language.

What, then, has Tarski actually shown us in showing that the idea of something being *true of* something, the general form of the "relation" of satisfaction, is the *central* idea that allows us to understand the structure of predication? In fact, the outlines of a partial answer to this question can be found in Heidegger's *disclosive* account of truth as founded on the most basic underlying "as"-structure of interpretive understanding. In particular, if Heidegger is right, the phenomenon that is ultimately at the basis of the possibility of *any* linguistic predicates being "true of" their objects (as well as any other phenomenon of disclosure) is the phenomenon of unconcealment, itself founded on the existential-hermeneutic "as."

In a passage from *The Basic Problems of Phenomenology*, Heidegger specifies how the hermeneutic "as" can be understood as the underlying basis for the "is" of predication, and indeed for the structure of linguistic predication generally, in light of the general conception of truth as disclosure or (here) unveiling:

> So far as the "is" in assertion is understood and spoken, it already signifies [*bedeutet*] intrinsically the being of a being which is asserted about *as unveiled* [*als enthülltes*]. In the uttering of the assertion, that is to say, in the uttering of exhibition [*der Aufzeigung*], this exhibition, as intentionally unveiling comportment, expresses itself [*spricht . . . aus*] about that to which it refers. By its essential nature, that which is referred to is unveiled. So far as this unveiling comportment expresses itself about the entity it refers to and determines this being in its being, the unveiledness of that which is spoken of is *eo ipso* co-intended [*mitgemeint*]. The moment of unveiledness is implied in [*liegt im*] the concept of the being which is meant in the assertion [*des in der Aussage gemeinten Seins selbst*]. When I say "A is B," I mean not only the being-B of A but also the being-B of A as unveiled . . . The extant [*vorhandene*] entity itself is in a certain way true, not as intrinsically [*an sich*] extant, but as what is uncovered in the assertion [*in der Aussage entdecktes*].[29]

On Heidegger's account, it is therefore the basic way in which an entity, in being uncovered, is shown *as* something that provides the ultimate phenomenological basis for the explicit "is" of predication, or indeed the inexplicit structure of predication, when no form of the predicative "to be" is present in the sentence. It is thereby possible to see the predicative structure as phenomenologically grounded in the more basic "as"-structure, which is in a fundamental way *not* relational or synthetic with respect to the matters unveiled, but rather exhibitive of them. On the one hand, this can be seen as providing a concrete basis for the claim, basic to Davidson's account of the most important reason why Tarski's account of predication succeeds where all others have failed: namely that it (almost uniquely) refuses to treat the structure of a simple predicative sentence logically as the *synthesis* of two or more separate and individually referential elements. On the other, it extends the nonsynthetic form of the solution to the deeper underlying structure of all (truth-evaluable) phenomena of disclosure, whether or not they yield linguistic assertions at all.

Returning to the connection between satisfaction and truth, what kind of illumination of the notion of being "true of" does this suggestion provide? As we have seen, it illuminates conditions for the possible availability of entities to be *understood*, conditions that must be seen as aspects of their sense. And it is plausibly requisite to any full picture of predication that we give some such account. Otherwise it would be just mysterious how we should conceive of entities as *becoming* possible subjects for true or false predication, or what "transcendental" conditions must be met in order for them to be able to be the subject of true or false assertions by the speakers of a particular language. These are issues on which Davidson's own account of the basis of predication, as far as it goes, is simply silent. But as we have seen, they cannot be avoided in the context of a broader consideration of the phenomenon of transcendental truth that he invokes. Heidegger's picture, by contrast, gives a phenomenologically motivated basic clarification of them. In so doing, it *a fortiori* also speaks to other questions about predication, some of which are not, and cannot be, answered by Tarski's structure of language-specific truth-definitions alone. As we shall see in the next chapter, these include questions about diachronic changes in languages over time, as well as the extensibility of the notion of truth across *all* languages with a minimal kind of predicative structure.

If such a hybrid conception of the foundations of truth is to be defended, though, it will be necessary first to overcome the persistent impression that its two main elements—the logical-semantic foundation for the structure of linguistic truth in terms of Tarski's apparatus and the ontological foundation for the givenness of entities in terms of Heidegger's conception of truth as unconcealment—are simply and irreconcilably opposed to one another.

Commentary on Heidegger has sometimes made just this suggestion. On the one hand, some recent scholarship has emphasized the ways in which Heidegger's own commitments about truth appear to embody a critical position with respect to mainstream analytic theories. For example, Daniel Dahlstrom has recently argued that a major basis for Heidegger's positive views is his critique of a "logical prejudice" broadly characteristic, according to Dahlstrom, of the analytic tradition of Frege, Quine, and Davidson. The prejudice is that the assertoric or declarative linguistic proposition or sentence is the basic locus or most characteristic bearer of truth.[30] But on the other hand, from a position methodologically influenced by the "language-analytic" tradition, Ernst Tugendhat in the 1960s articulated a much-discussed critique of Heidegger's conception of truth in his *Der Wahrheitsbegriff bei Husserl und Heidegger*, and in a related shorter article, "Heidegger's Idea of Truth."[31] In both pieces Tugendhat argues that, in grounding truth in the disclosure of entities, Heidegger's conception cannot account for important specific features of the phenomenon. In particular, Heidegger's account understands truth basically as a phenomenal *event* of unconcealment that either occurs or does not occur. But such an account must fail, Tugendhat argues, to distinguish between a "true" or accurate event of appearance and a "false" or inaccurate one. With this, he suggests, the basic difference between truth and falsity itself is abandoned, and the hope of providing an adequate characterization of sentential truth is lost.

According to the line of critique Tugendhat develops in detail in *Der Wahrheitsbegriff*, this has several other problematic consequences for Heidegger's picture of sentential truth. One is that it appears to make truth dependent on *ontic* events of discovery: thus, for example, if truth is grounded in the disclosure of entities, it appears to follow that a claim about an entity, for instance a newly discovered astronomical phenomenon, is not true until it is discovered to be. For similar reasons, Heidegger's account seems to have trouble accommodating sentences or propositions that are understood but not yet verified: such a sentence would seem indeed to disclose or reveal the entities treated by it, but does not seem to be reasonably treated as *true* simply by virtue of doing so.[32] Second, Tugendhat suggests that Heidegger's account has difficulty with *relational* sentences: for example, it is not at all clear from this account what we are to do with even a simple relational sentence such as "Socrates is older than Plato." Should we think of the truth of this as grounded in the unconcealment only of Socrates (since he is the grammatical subject)? But then we must think of his being unconcealed in terms of a relational property (being older than Plato) rather than as his being unconcealed simply as he is in himself. This would threaten to make all relations into internal properties of an individual, and since each object is related somehow or other to all others, it would imply that the full unconcealment of an individual object also discloses the whole universe.[33] Third and finally, as Tugendhat suggests in passing, an even harder case is that of (true) negative judgments of existence, for instance the judgment "Santa Claus does not exist" or "There are no unicorns."[34] It is not at all clear how the truth of these judgments can be grounded in the "disclosure" (uncovering, etc.) of the entities mentioned, since these entities do not even exist.[35] Here, as Tugendhat suggests, it is accordingly unclear what it could mean to speak of the "thing itself" or of the true proposition as disclosing it "as it itself is."

Tugendhat further suggests that *only* a conception that takes the truth of sentences or sentence-like structures as basic can account for truth in its close connection to predication.[36] It is accordingly necessary, in accounting for these basic logical features, to begin with the articulated structure of a language already in view, and there is no hope for an independent ontological grounding of truth in terms of essentially prelinguistic or nonlinguistic properties of entities. By apparent contrast with this, Heidegger himself, at least on Dahlstrom's reading, maintains that such a privileging of linguistic structure, and in particular the predicative structure of the assertoric or declarative sentence, can only be the expression of a "logical" prejudice which is coeval with the metaphysical interpretation of being as objective presence and representation, and is to be overcome by means of an ontologically and pragmatically deeper analysis. For both lines of interpretation, Heidegger's account of truth as unconcealment would thus seem to be deeply and fundamentally incompatible with the analysis of linguistic-logical predication, of the type that Davidson presupposes and Tarski (among others) carries out.

However, the impression of incompatibility can be largely overcome, in both directions, by bringing out the specific structure of the hermeneutic "as," which is, for Heidegger, at the ontological basis of the phenomenon of truth. In particular, by seeing how the hermeneutic "as"-structure both *underlies* the structure of predication in the assertion *and* has a deeper ontological ground in the presentation of beings in essential relation to their ways of being, we can see how it accounts for the phenomenon of propositional truth in terms of what is genuinely the unconcealment of entities, while *nevertheless* also capturing the specific logical-linguistic features to which Tugendhat points as essential.

As we have seen, Heidegger understands the "existential-hermeneutic" "as"-structure, whereby something is presented *as* something, as the most basic structure underlying any possible understanding of entities. The structure is further articulated into the three structural moments of *fore-having* [*Vorhabe*], *fore-sight* [*Vorsicht*], and *fore-conception* [*Vorgriff*].[37] As thus articulated, the underlying hermeneutic "as"-structure has several logical features relevant to the present question.

First, although the structure *underlies* the possibility of sentential predication, it is more *general* and *basic* (in one sense of "basic") than this structure itself. In particular, it is operative already in *various* kinds of circumstances of uncovering that possess a dimension of "veridicality" or truth-evaluability, including perception and engaged practice, whether or not there is any explicit conceptual articulation of a judgment or linguistic articulation of a sentence.

Second, the structure is nevertheless fully (and irreducibly) *intensional.* That is, if, in accordance with the structure, an item is uncovered as having a certain trait or characteristic, this does not generally imply that the substitution of coreferential terms would preserve the truth of the statement that the item is thus uncovered (for instance, if a particular apparent celestial object is uncovered *as* the morning star, there is no implication that it is thereby *also* uncovered as the planet Venus). In this sense, what is unconcealed through the basic "as"-structure is always "under" a "mode of presentation," though we should not think of this as equivalent to being "under a conception" (since there need not be any *explicit* conception at all).

Third, the structure is irreducibly *holistic*: the uncovering of a particular entity as being a particular way in general depends on the whole surrounding structure of its relations of significance to other entities. In the context of everyday concern, this holism has its basis in what Heidegger characterizes as the "in-order-to" structure of ready-to-hand "equipment." These items of everyday use do not show themselves simply as themselves, but always in relation to other items and purposes, in relation to which they comprise a total system of "references" or holistic significance.[38] Under the condition of interpretation, this holistic context of significant references is "taken apart" in its in-order-to structure in such a way that its "as"-structure can be brought out explicitly. The articulation of this "as"-structure of things also brings out more explicitly the "totality of involvements" that the ready-to-hand always, as such, already involves. The interpretation is always thus grounded in a totality of involvements that we already "have" in advance of its explicit development. This is the "fore-having" that amounts, according to Heidegger, in an "always already prior disclosure of world."[39]

These features of the "as"-structure—especially the last two (intensionality and holism)—may seem prima facie mysterious, given that this structure is not here seen as imposed or primarily produced by a subject's cognitive or representational activity. How is it possible for this structure, we might wonder, to be irreducibly intensional and holistic, if these features are indeed not to be seen as the outcome of a subjectively imposed framework, but rather are rooted in characteristics of the things themselves as disclosed in relation to their world as such? As we shall see in more detail below, however, the answer to this question lies in the way in which the hermeneutic "as"-structure is itself related to the structure of disclosive truth, and *thereby* to the ontological difference between being and beings. Indeed, if, as I have suggested here, sense is itself understood not primarily in terms of subjective cognition or conception but rather construed, in an ontologically realist way, as *mode of presentation,* there is no evident obstacle to seeing an irreducible and basic intensionality as characteristic of *any* possible presentation—that is, any possible (way of) *being*—of an entity as such. And because of the way the conditions of presentation are themselves essentially related to the *total* structure that makes entities as such intelligible in one way or another, this intensional structure is also irreducibly *holistic*, in the sense sketched in the last paragraph. Understood in ontological terms, the basis of both aspects of the hermeneutic "as," as well as the source of the reality of intensionality in the world (and hence the basis of any possible process of subjective representation, *as well as* the pre- or nonrepresentational meaningfulness of things) is the excess of being and the presentation of beings *over* beings themselves that is implicated in the ontological difference. And as a result of this, the basic hermeneutic "as"-structure can be seen as the underlying condition for any representation of beings, but also equally for their *being* the (intelligible) ways that they are to begin with.

What, then, of the broader methodological dispute that gives both Dahlstrom's interpretation, and Tugendhat's criticism, their distinct conceptual and motivational points? Here it is important to consider that, as Heidegger himself suggests with respect to judgment, assertoric or sentential truth may be a phenomenon with "more than one kind of foundation."[40] More specifically, it is not at all obvious that an *ontological* foundation in disclosure and in the more ultimate structure of the "as" (which indeed characterizes, as we have seen, the "being of entities") is not compatible with a different kind of structural *semantic* foundation for the truth of sentences in a language, one that comes into view much more clearly through accounts such as Davidson's. To see this, it suffices to note the obvious sense in which, on the one hand, the truth or falsity of sentences depends on the ways of *being* of the entities involved in them, whereas, on the other, particular entities are only intelligibly available *as* entities, *even in* unthematic *praxis* and everyday dealing, through and (partially) because of a language which yields terms for their consideration and description.

On this kind of picture, there is thus a twofold "grounding" of truth—understood as *both* the truth of sentences *and* the "availability" of entities—having both a "logical" and an "ontological" dimension. This twofold grounding is not simply "ontic," in either of its dimensions. It does not, for instance, point basically to the dependence of one entity or kind of entity on another, or to the empirical dependence of one kind of sentence structure on another in the representation or cognition of a particular natural language. Nor does it establish anything like ontic *conditions of possibility* (e.g., causal ones) for the existence of any entity or the taking place of any event. It is, rather, *irreducibly* ontic-ontological: it is grounding across the ontological difference, and is thereby grounding not simply of entities and sentences but also of their *sense* or intelligibility as this arises from the twofold structure of their truth.[41]

How, though, should we understand such (ontic-)ontological *grounding* itself? Throughout the discussion so far, I have described (and will continue to describe) various phenomena as more "basic," foundational, original (etc.) than others, or as ontologically "prior" to them. This kind of talk obviously invokes an order of priority that must itself be motivated in the terms of the ontological problematic, that is, the question of the sense and truth of being, and hence in terms of the ontological difference. Heidegger's own understanding of ontological or phenomenological "grounding" and "founding" develops in part from Husserl's. For Husserl, founding [*Fundierung*] is a relation of ontological *dependence*: A is said to found B if B is dependent for its *existence* on A. However, since the ontology of founding is carried out wholly within the broader project of transcendental phenomenology, the dependence-structure of foundation applies not only to spatiotemporal objectivities, but just as well (and even more centrally, in fact) to senses and contents. As such, it has the significance not only of (what we might today call) metaphysical or ontological dependence, but also that of (what we might see as) formal-epistemological foundation. What is at issue in the question of the foundation of a specific sense or category is not only the ontological status of entities falling under it, but also the basis for the *manifestation* or possible *presentation* of that sense or category itself.

In the development of the properly ontological problematic of being and beings—the problematic of the ontological difference—Husserl's jointly metaphysical and formal-epistemological idea of the grounding of the categories is replaced by the hermeneutic structure of formal indication as the interpretation of facticity. With this, ontological grounding is thought in terms of the ontological difference and as operating across it: that beings are grounded in "something" else—their being—means that this—their being—is "responsible" for their presence and presentation *as* the kinds of beings they are.[42] That *A ontologically grounds B* then has, in general, the meaning that A is responsible for the *presencing* or *presentation*—the making-present—of B as such.[43] But it would not be sufficient to say simply that being *makes* entities present or *causes them* to be present. For any such assertion of a concrete and active relationship between being and entities takes the form of an ontic assertion about the relationship of entities, and being is not an entity. Rather, being *grounds* entities in that it *opens* (hence, dis-closes) *a domain* for their presencing.[44] As such, it grounds *both* their existence *and* their sense. But beings are made present in a variety of ways. Some of these ways are directly linguistic: linguistic judgment, for example, is one kind of presencing. This kind of presencing is obviously conditioned by the specific structure of the language in which it takes place. But some are not, or not as directly so: there is, at any rate, no reason why an entity cannot (as Heidegger indeed emphasizes in *Being and Time*) show up as *having* a specific kind of character, understood in terms of its distinctive "in-order-to" relationality to other entities, without being the subject of any specifically *linguistic* judgment or predication.

3. Consequences of "Transcendental" Truth 1: Sense and Presentation

I have argued for a hybrid understanding of transcendental truth on which it is has both a linguistic-hermeneutic dimension, articulated through the employment of Tarskian truth-theories in radical interpretation, and an ontological-hermeneutic one, articulated as interpretive disclosure. In this section and the next, I shall briefly develop some implications of this twofold understanding for two further questions. The first is the question of the structure and existence of *senses,* understood (following Frege) as modes of presentation. The second is about the semantical and dynamic behavior of languages (such as English) which contain both their own truth-predicate and the general possibility of forming names for their own sentences (for instance, by means of quotation or some similar device).

Frege often describes the sense of an expression, whether it be a name, a concept-term, or a sentence, as a "mode of presentation" (*Art des Gegebenseins*) of a referent. His most basic reason for doing so is the same as his reason for introducing the distinction between sense and referent to begin with: the need to account for the *informativeness* of judgments of identity.[45] On the conception, it is necessary in order to account for this informativeness that the same referent (e.g., a triangle or the planet Venus) can be presented or given in any of various different ways, and that it *is* the same object may not be evident from these presentations alone. The sense of a name is thus to be understood as a way in which its referent is presented or given, and, analogously, the sense of a sentence is understood as a way of presenting its truth-value.

As commentators have noted, this conception of senses as modes of presentation is in some tension with another picture Frege sometimes employs, particularly with respect to sentential senses: namely that of senses as nonphysical *entities* of a special kind, capable of being grasped in thought, and possibly existing in a "third realm" beyond those of the spatiotemporal and the individual-subjective.[46] One particular place in which the tension shows up is in Frege's own account of indirect discourse or *oratio obliqua.* On Frege's account, a sentence in oblique discourse (such as "The planetary orbits are circles" as it figures in "Copernicus asserts that the planetary orbits are circles") has an oblique or "indirect" referent which is not its ordinary one (i.e., a truth-value) but is rather its (ordinary) sense.[47] In terms of Frege's overarching picture of the distinction between concepts and objects, this means that senses, in oblique contexts, are treated as particular kinds of *objects.* But it is not clear how to individuate or distinguish senses if we treat them as objects in this way. In particular, senses cannot be individuated simply by the words used to express them; for it appears correct to describe Copernicus (continuing the example) as having asserted the same thing as one would today assert with the words "The planetary orbits are circles," even though he never spoke those (or any other) English words. On the other hand, we also cannot identify the senses of two terms, even if their referents are *necessarily* identical (e.g., we cannot identify the sense of "equilateral triangle" with that of "equiangular triangle"), since the judgment of identity may still, in such a case, be informative.

In *Meaning and Necessity,* Carnap argues that Frege's treatment of sense, reference, and indirect discourse points toward a more general antinomy which arises for any theory that treats the various types of expressions composing a sentence as each having a distinct sense and reference.[48] Specifically, if we allow the general substitution of extensionally equivalent or even just necessarily coreferential terms into sentential contexts, including sentences containing oblique contexts, the result will be contradictions in the truth-values assigned to some of these sentences.[49] Frege's device of ascribing as the reference of a sentence in *oratio obliqua* its (ordinary) sense can be seen as a way of avoiding the antinomy, but it leads, as Carnap notes, to a kind of infinite proliferation of entities. If a particular sentence is used obliquely, its sense is thereby named; but then the name must be conceived as different from the sentence in its usual use (which, instead, names a truth-value rather than a sense). This name then has a further sense, which itself must have a further name, and so forth. The treatment of senses as possible objects of reference, which Frege introduces to attempt to deal with the special problems of indirect discourse, thus necessarily introduces an infinite series of names and special objects for each sentence that can appear embedded in such discourse.

As an alternative, Carnap develops what he calls the "method of extension and intension." On this method, intensions or senses are defined by the semantic rules laid down in advance for "the use of corresponding expressions in language systems to be constructed," and substitution of coreferring terms is not generally permitted into intensional contexts.[50] Carnap observes that any picture committed to treating linguistic expressions generally as names (or as always having referents, in Frege's terminology) will lead to the antinomy of the name relation and invite, by invoking senses (or intensions) which must be nameable, something like the infinite proliferation of names and entities that Frege's picture involves.

What Carnap does not point out is that his own method of extension and intension, while prohibiting the substitution of coreferring terms in intensional contexts internal to a language, nevertheless leads to a formally similar problem on the level of the determination of a language itself. For Carnap, remember, the rules that determine the analytic truths (or L-truths) of a language and its logical equivalences (or L-equivalences) are to be stipulated in advance in such a way as to explicate the preexisting notion of necessity or analyticity.[51] But to begin with, this does not solve the problem Frege's invocation of the sense/reference distinction was originally meant to answer, since Carnap *does* permit the substitution of (at least) *L-equivalent* expressions in *all* (intensional as well as extensional) contexts. If, for instance, the axioms of geometry are considered to be among the semantic rules definitive of the linguistic system, then Carnap's system will allow "X believes the triangular figure is equilateral" to be replaced with "X believes the triangular figure is equiangular," and the informativeness of the identity judgment "An equilateral triangle is equiangular" is not explained.[52] Furthermore, and more generally, if the semantic rules determining L-truth and L-equivalence are themselves conceived as explicating preexisting determinations of analyticity and logical equivalence in a natural language, the question arises: on what basis can these preexisting determinations themselves be justified? The stipulation of semantic rules *explicative* of analyticity in a natural language as spoken would thus seem to presuppose a further explicit stipulation of the rules already *constitutive* of the natural language in question. But it is impossible to suppose that this stipulation of rules could take place, unless the rules in question were *already* in place.

This argument against Carnap's conventionalist picture of the institution of linguistic rules is made explicitly by Quine in "Truth by Convention," and later formulated somewhat differently in the classic "Two Dogmas of Empiricism."[53] It appears to affect any picture on which the intensional structure of a language is considered capable of being wholly and uniquely described from outside that language itself, whether the object of description is understood as a corpus of explicitly stated rules, conventions of usage, or even just regularities implicit in practice.[54] The aporia that results is, in the context of the present discussion, a direct consequence of treating the intensional structure of a language as capable of objective description from without. As such, it may be understood as a consequence of "treating senses as objects" (in a suitably extended sense of "sense" and "object"). With respect to the origin and constitution of natural (historical) languages, it appears to demonstrate the untenability of any conception of their intensional structure as conventionally or stipulatively instituted, and thereby point to a deeper problematic of their historicity.[55]

At any rate, there are good reasons to think that Frege's own motivations in introducing the sense/reference distinction to begin with should have led him to try to avoid these interrelated problems, even if he was not completely successful in doing so. According to Dummett, the common complaint that Frege does not tell us much about what constitutes senses or how to individuate them is only partly justified; for Frege has, after all, specified the sense of an expression as "the manner in which we determine its reference," and has furthermore said much about the different kinds of referents of different types of expressions.[56] Nevertheless, Dummett admits that Frege's usual way of indicating the senses of words and symbols is not to describe these senses directly, but rather just to state the reference. The reason for this procedure, according to Dummett, is that, since the sense of an expression is just the mode of presentation of a referent, we cannot expect to be able, in general, directly to specify senses, for instance by a pronouncement of the form "The sense is . . ." Thus, at least on the "healthier" of the two strands of Frege's notion of sense (Dummett means, here, the strand according to which senses are modes of presentation rather than ideal objects) "the sense of a proper name is the way we arrive at the object, but not conceived as a means to a separable end"; sense is thus "better understood as the manner in which we pick out the object than as the route we take to it."[57] Instead of describing senses directly, which we thus cannot do in general, according to Dummett, when we wish to convey or stipulate a sense we may adopt the procedure that Frege does adopt, which is to "choose that means of stating what the referent is which displays the sense."[58] Dummett here appeals to the early Wittgenstein's distinction between saying and showing, suggesting that in thus *saying* what the referent of an expression is (in a particular way) we may succeed in *showing* its sense.

As Dummett acknowledges, however, this leaves open the question of *what is actually known* in *knowing the sense* of an expression of a certain type.[59] The distinction between sense and reference was introduced, after all, to track salient distinctions in knowledge, and in particular to account for the possibility of gaining knowledge of the truth of a statement of identity, so it seems reasonable to ask this further question. One possibility that Dummett considers is that "to know the sense of a proper name is to have a criterion for deciding, for any given object, whether or not it is the bearer (referent) of that name" and similarly that to know the sense of a predicate or relational expression is to have a criterion for deciding "for any given object[s]" whether the predicate or relational expression applies to them.[60] On this sort of view, to grasp a word's sense is thus to possess a kind of *ability* to determine the truth-conditions of sentences involving it, and this ability is further to be understood, in the case of names, as the ability to recognize the requisite objects (or recognize whatever counts as establishing that the name applies to a particular object).

The trouble with this, as Dummett notes, is that it is not in fact legitimate, in the context of a full picture of sense, simply to characterize the knowledge of sense as a matter of the ability to recognize "a given object." For

> in understanding a proper name or predicate, I am supposed to be able to recognize something as establishing that a given object is the referent of the name or that the predicate applies to it: but *what* is it that I recognize to be established? That such-and-such a name stands for the object, or that such-and-such a predicate applies to the object—indeed: but *which* object? The given object, of course: but here we have a right to ask, "How was it given?"[61]

In fact, as Dummett points out, the conception of knowledge of sense as simply consisting in an ability to recognize objects cannot succeed, since an object cannot be recognized at all unless it is first presented *in some specific way or other*, and this already involves (in terms of the conception of senses as modes of presentation) that it is presented by means of *some sense or other*. "We are," Dummett concludes in a partially parallel discussion, "never given an object, complete in itself; we can think about it, speak of it or apprehend it only as presented to us in some particular way."[62] Thus the conception of knowledge of senses as consisting in the ability to recognize objects as the bearers of names, or as falling under predicates, cannot succeed, except perhaps in the context of a much broader conception of sense. Such a conception would have to be inclusive of the observation that the grasping of a sense (for instance) of a proper name must include the ability to grasp objects as *of* a type or category, including grasping general "criteria of identity" for objects of that type.[63] For instance, even the use of a proper name in connection with a demonstrative gesture in expressing a judgment of recognition ("This is Fido") already involves the availability of sortal criteria for objects of the type (e.g., "dog") that are not wholly given in the recognitional judgment itself.

It is here that we may appeal to Heidegger's specific disclosive conception of truth as providing an appropriate conception of sense as mode of presentation that has undergone the requisite broadening with respect to Frege's own conception. As we have seen, Heidegger's conception of truth as disclosure sees it as ontologically grounded in the primary structure of the hermeneutic-existential "as," or the presentation of something *as* something. Sense is itself, for Heidegger, grounded in the "projective" phenomenon of interpretation characterized by the "as"-structure, and so it is, on this conception, nothing other than "mode of presentation" in a generalized form. As we saw above, the underlying existential/hermeneutic "as"-structure is inherently intensional and plausibly *preconditions* (without being reducible to) narrower intensional and "intentional" phenomena, such as propositional and intentional attitudes, by pointing toward the general hermeneutic conditions for *any* intentional "accessibility." On Heidegger's picture, it is thus possible to agree with Dummett's point, that we can recognize, speak about, or apprehend objects "only as presented in some particular way" (i.e., including that they are presented as being of some particular type) and, as well, to give an expanded and phenomenologically motivated account of how this presentation occurs. This account, understood as a suitable generalization of Frege's narrower conception of sense as mode of presentation in language, connects that conception back to the phenomenological and ontological ground of possible givenness of objects. It is this ground, as we have seen, that must plausibly *underlie* any account (such as Frege and Dummett attempt to give) of how linguistic signs can function to express senses that are themselves conceived as modes of presentation of objects.[64]

Most significantly, however, the supplementation of the broad picture of truth with the Heideggerian conception of truth as disclosure allows for an *ontological* clarification of the status of sense (conceived broadly as mode of presentation) that indicates the actual ontological reasons for the "antinomy of the name-relation" and the related aporias that arise for pictures that treat senses as objects to which reference can be made. As we have seen, on the Heideggerian picture, sense, in accordance with the underlying hermeneutic-existential "as," is always related to presentation, and presentation is as such constitutively linked to being in that the underlying "as"-structure always reveals entities in their being or otherwise. It is thus that sense, while it typically and primarily characterizes entities, is also always structurally linked to their being, and also to their unconcealment within the ontic-ontological structure of Dasein.

A general reason for concluding that modes of presentation cannot be treated as objects, then, can be found in the observation that sense is always characterized by a twofold structure bearing a twofold indicative meaning, directed not only to entities but *also* to their being; and because of the ontological difference, being itself is not an entity. Modes of presentation, in Heidegger's sense, cannot therefore be characterized simply as entities, and cannot be accounted for wholly in terms of the (merely ontic) relationship between two or more beings. Instead, they require an accounting (at least in part) in terms of the ontic-ontological difference itself.[65]

With this in mind, it is possible to consider the implications of Heidegger's broadened picture for the narrower issue specifically treated by Frege, that of the senses of specifically linguistic expressions as modes of presentation (of referents). From the broadened Heideggerian perspective, as we have seen, linguistic senses can still be treated, as Frege does treat them, as given by whatever allows *truth-conditions* for the sentences of a language as a whole to be determined. But this determination must, in addition to the linguistic-structural determination of the truth-values of sentences from "primitive" relations of satisfaction (in accordance with a Tarskian truth theory), *also* include whatever determines the presentation of beings (in such a way that they *can* plausibly figure in relations of satisfaction, or other reference-like relations with linguistic terms, to begin with). Dummett actually grasps this point, at least in a negative way, as is evident in his recognition of the need to address the ways in which objects are given as part of a general theory of sense. But although he also recognizes that modes of presentation cannot be treated as separable objects of reference, on pain of aporia and infinite regress, he is led by the apparent demand to explain what knowledge of a linguistic sense consists in to attempt to countenance such knowledge as possession of an ability to recognize objects, or more broadly to use linguistic terms meaningfully in specific, epistemically characterizable ways.

The attempt to describe knowledge of senses in terms of recognitional abilities does not succeed, since as Dummett also recognizes, the description of knowledge of sense in terms of abilities to recognize objects itself presupposes the availability of the senses it is supposed to explain. But while recognizing this failure in the case of the narrow recognitional theory, Dummett nevertheless persists in supposing that it *must* be possible to characterize the knowledge involved in knowing a linguistic sense noncircularly in terms of *some* notion of epistemic ability, ultimately at the basis of our ability to identify something *as* being the case when it is. This is what yields Dummett's attempts to describe knowledge of sense in terms of such formulas as the "grasping of a rule," the possession of a specifiable ability, adoption of a method, the possession of specifiable criteria, or the knowledge of a means of "picking out" something from among others. But as we have seen, if it must (on pain of aporia) be impossible to treat senses in general as separable objects of reference, then it appears likely that these attempts to characterize sense noncircularly in terms of some prior available means must fail as well. Moreover, to introduce them as figuring centrally in explanations of what it is to grasp a sense, as Dummett does, is effectively to locate epistemic notions at the heart of the systematic theory of sense and truth, a strategy which is (as Dummett himself recognizes) in substantial tension with Frege's own realist approach, and which we have good reason to think cannot succeed in capturing the specific sense of truth, in any case.[66]

On the recommended Heideggerian alternative, there is no requirement, in general, for the knowledge involved in grasping the sense of an expression to be portrayed as consisting in the possession of any ability, adherence to any practice, or any relation of the individual subject simply to any other (ontic) event, object or process. This negative point is just the obverse of the positive recognition that sense as a *phenomenon* can only be described in terms that are ontological as well as ontic, and that this ontic-ontological double structure (itself inherent to the structure of Dasein) is essential to the specific phenomenon of truth itself. Significantly, this allows for a generalized discussion of linguistic sense that situates it as a specific phenomenon within the broader context of the ontological concept of presence and presentation as such. It thereby allows a broadened discussion, as well, of the *temporal* determinants of the possibility of presence in the broad as well as the narrower (i.e., linguistic) sense, a discussion in which the temporality of *language* is, necessarily, equally at issue.

From this perspective, the picture of an already-constituted language, determined and fixed as a total structure, coming to bear on things simply by confronting a world already constituted as a totality of determinate objects bearing determinate properties, is a complete fiction. Instead, it must be acknowledged that objects and their determinate properties and relations emerge, in a temporally *complicated* sense, only with the development of a language and the specific possibilities of expression and presentation that its structure, at any given point of its development, makes possible. That fictional picture of the language-world relation is, however, the one that ultimately motivates conceptions, such as Carnap's, on which a language can be conventionally instituted or stipulated as a whole simply by fixing its semantic rules. It is also the picture that may be thought to motivate Tarski's original idea of truth-predicates for artificial languages as definable in terms of "primitive" relations of satisfaction, though not (or at any rate, not in the same way) Davidson's modified picture of the interpretation of natural languages, which is inherently holistic and reconstructs satisfaction from truth rather than vice versa. From a perspective that takes seriously the ontological difference, however, such fictional pictures of the "language-world" relation must cede to one that sees linguistic sense as itself essentially articulated *along with* the linguistic articulation of the world on which it bears. This includes the determinate and variable possibilities a language affords, at a time, for the presentation of its objects and phenomena. With this, we can begin to see a phenomenological basis for challenging the Fregean assumption that senses must be timeless or eternal existences whose own temporality is quite independent of the temporality of the specific phenomenon they are actually invoked to explain, that of the presentation of objects. This is to open, in other words, the question of the underlying temporality of sense as (in the Heideggerian jargon) the question of the *being* of language, including the question of its temporal status, continued existence, and its possibilities of gradual or radical transformation or change in coarticulation with those of the world it discloses.[67]

4. Consequences of "Transcendental" Truth 2: Limit-Structure of Linguistic Truth

I turn, now, to another issue about linguistic truth on which the suggested Heideggerian extension of the concept of truth has specific bearing. This is the question of the structure of a language which contains, as natural languages generally do, its own truth-predicate (e.g., the English predicate "true," as characterizing sentences in English). In "The Semantic Conception of Truth," Tarski famously argued that any language that includes its own truth-predicate along with devices for forming arbitrary names for its own sentences (such as the device of naming sentences by quoting them, or some other suitable device) would, in connection with the structure of T-sentences, necessarily produce a contradiction.[68] For it is possible in any such language to construct a "liar" sentence of the form

L: "L" is not true

and, assuming the relevant truth-predicate is in the language under discussion, the T-schema for L produces the contradiction

"'L' is not true" is true iff "L" is not true

that is,

"L" is true iff "L" is not true.

Tarski accordingly suggested that no language that contains its own truth-predicate could be formally characterized in accordance with the T-schema; it was thus necessary to discuss and define the truth-predicate for each language under consideration in a *different* language that contains all the sentences of the original language (or translations of them) as a proper part. Given this, it is impossible to formulate the liar sentence, and the contradiction is avoided.

The procedure has some plausibility in the case of formal languages, for which there is a stronger metalanguage (such as English) readily available. But even in this case, one can reasonably wonder how the notion of truth in the metalanguage itself is defined or definable; here, the general application of Tarski's solution would involve the postulation of an indefinite open hierarchy of metalanguages, each stronger than the last, in which truth for each language is definable only by the next higher one. And if we are interested primarily in natural languages rather than formal ones, the device of defining truth in a metalanguage is of no use to begin with. For these languages plausibly contain their own truth-predicates, and there is not, in general, a "stronger" metalanguage available that can express everything the initial language can express and more. Indeed, it is plausible, as Tarski himself recognized, that every natural language has (as such) "universal" expressive power, and is therefore not to be modeled without contradiction by the T-schema. Tarski himself concludes that it is therefore hopeless to attempt to use the T-schema to characterize the structure of the truth-predicates of natural languages, and accordingly that these languages are, in a basic sense, unformalizable.

More recently, though, some philosophers have considered how the structure of truth can be formally treated in accordance with the T-schema even for those languages which, like English, contain their own truth-predicates (as well as the resources to make descriptive reference to their own sentences). In his influential "Outline of a Theory of Truth," Kripke notes that Tarski's hierarchical approach cannot handle situations that occur routinely in natural languages where the truth of sentences is discussed, such as the situation in which Dean says, "Most of Nixon's statements are false" and Nixon says, "Most of Dean's statements are false."[69] As Kripke notes, there need not be any actual contradiction here, since both Dean's statement and Nixon's statement can be true without producing a contradiction; but Tarski's hierarchical device implies both that the truth-predicate Dean uses in describing Nixon's remarks must be at a higher linguistic level than the one Nixon uses and that (conversely) Nixon's must be at a higher level than Dean's. Kripke suggests, accordingly, that Tarski's account in terms of a hierarchy of languages and different truth-predicates should be replaced by one on which sentences, rather than being assigned to fixed levels of a linguistic hierarchy, are allowed to "find their own levels" within a language conceived as unitary. In particular, given an initial partial interpretation which assigns truth-values to some of the language's sentences, sentences involving ascribing the truth-predicate to other sentences can be assigned truth-values at a higher level than the sentences discussed. The hierarchical construction can be iterated arbitrarily and even through transfinite ordinal levels. Kripke shows that, within the iteration, there will be certain *fixed points* at which some sentences attain a stable value (such that, if true at that level, they will be true at all subsequent levels of the hierarchy, or if false there, they will always subsequently be false), and these can then be treated as determinately true or false. Nevertheless, there will still be some sentences (such as the liar sentence itself) that never attain a stable truth-value on any level; these are treated as "ungrounded" and as exhibiting truth-value gaps—that is, as being neither true nor false.

Drawing on the implications of Kripke's "fixed-point" construction but modifying its intuitive basis, Gupta and Belnap have proposed a "revision theory" of truth that attempts to explicate the concept of truth as a *circularly* defined one. In particular, whereas Kripke understands the stages of the hierarchical construction as iterated interpretations of a language at ever-higher inductive levels, Gupta and Belnap understand the Tarski biconditionals as pointing toward an underlying concept of truth that is circular in the sense that the *definiendum* (truth) is in some cases contained in the *definiens*.[70] Such a definition, as Gupta and Belnap note, does not permit the determination of the extension of the definiendum noncircularly, but it may nevertheless give a rule according to which we can determine the *final* extension of the definiendum given an *initial* hypothesis about its extension.[71] The reinterpretations of the truth-predicate that occur on different levels of the hierarchical construction, including (some) fixed points, are thus interpreted as *revisions* of a truth-predicate that is defined, in itself, circularly. This kind of circularity is, Gupta and Belnap hold, at the root of both the irremediably paradoxical behavior of some sentences (such as the Liar sentence) and the fact that others eventually attain stability at some fixed points.

Against both pictures (and others), Graham Priest has argued that taking seriously the Tarskian T-schema in application to the truth-predicates of natural languages such as English requires recognizing that the concept of truth they formulate is inherently contradictory in the sense that such a language will contain true contradictions.[72] Attempts to avoid a contradictory semantics while preserving the Tarskian schema, such as Tarski's own as well as those of Kripke, Gupta, and Belnap, are therefore to be rejected. Priest's main argument against theories such as Kripke's, which invoke truth-value gaps and deny the general applicability of the law of the excluded middle, is that no such solution can, in fact, avoid contradiction entirely. For instance, for a sentence, *a*, that turns out to be ungrounded in Kripke's sense and is thus treated as lacking a truth-value, "*a* is not true" is nevertheless itself true (since sentences that lack a truth-value are not true). Thus we can consider the "extended" liar paradox:

S: "S" is not (stably) true.

If "S" is true, it is not (by the T-schema), and if it is false *or* valueless (for instance if it is "ungrounded" in Kripke's sense), then it is true. Accordingly, even the hierarchical construction in terms of stable truths at fixed points fails to achieve an unparadoxical classification of sentences across the language as a whole.

As Priest shows, it is a consequence of the liar phenomenon in its extended version that a Davidsonian truth theory for a natural language containing a truth-predicate cannot be consistently decidable, or even finitely axiomatized. This remains the case *even if* paradoxical sentences are excluded from it as meaningless (in that they lack truth-values) and hence beyond the scope of the applicability of the T-schema to produce true T-sentences.[73] For even in this case, on the assumption that there is a recursively enumerable (hence axiomatizable) theory that can prove (all and only) the T-sentences for *meaningful* sentences, it is possible to derive a contradiction of the "extended liar" type. In particular, it is possible to formulate a sentence β saying of itself that it is *either* not true *or* not meaningful. Supposing this sentence is meaningful, the T-schema must hold for it; thus if it is meaningful, it is true iff it is either not true or not meaningful. Thus it is not meaningful. Therefore it is not meaningful or not true, that is, β. But if β is true, then it is meaningful (since all true sentences are meaningful). This is a contradiction, and so it follows that the initial assumption is false: there can be no recursively enumerable theory of this type. Priest concludes that it is impossible to block the paradoxes by excluding contradictory sentences from a truth theory while maintaining its axiomatizability. Any theory of meaning for a language containing its own truth-predicate (and the minimal expressive resources needed to express Peano arithmetic) will thus be either contradictory or incapable of recursive axiomatization, or both.

Priest gives a further argument against Gupta and Belnap's revision theory specifically, this time bearing on the structure of the underlying basis of the "conventions" that provide for the definition of truth, on their account.[74] As Priest notes, since the revision theory identifies the truth-value of sentences with their truth-values on stabilized interpretations, it can regard the T-scheme as holding only for stably true or false sentences, rejecting its applicability to paradoxical sentences (such as the liar). But Priest questions what this is supposed to show: in particular, how does the elaborate formal construction involved in the distinction between stable and unstable sentences actually relate to the semantics of English as spoken? The construction involves, after all, a hierarchy of interpretations that must be continued not only through finite but also through transfinite ordinal levels. If this hierarchy of interpretations is supposed to explicate the meaning of the truth-predicate of an actually spoken language such as English, it is thus apparently necessary to attribute to the individual speakers of the language at least an implicit grasp of the (highly complex) notions of a transfinite ordinal, of transfinite induction, and so on. But that ordinary speakers of English generally grasp these notions in using the ordinary notion of truth seems highly implausible. Similarly, Priest questions how the successive stages of revision are actually to be interpreted: it seems inappropriate to interpret them as actually temporally or chronologically ordered (so that the extension of the truth-predicate would be revised periodically, perhaps every week or every month), but it is unclear how else to interpret the claim that truth is successively "revised," as Gupta and Belnap claim.[75] Finally, even if we follow Gupta and Belnap in taking sentences to have the semantical properties they do when the revisions finally stabilize, there will still be the possibility of generating paradoxes of the "extended liar" type anyway.[76]

All of these constructions and interpretations of the significance of the paradoxes are relevant in the current context because they attempt formally to explicate what I have here called the "transcendental" concept of truth, which is plausibly the concept underlying truth-predicates of languages such as English that contain their own truth-predicate and that possess "universal" expressive power in Tarski's sense. As we have seen, any attempt seriously to consider the structure of this transcendental concept of truth requires rejecting Tarski's hierarchical structure of languages with distinct truth-predicates, since none of these captures the target notion under consideration. The transcendental concept of truth, as I have discussed it here, is itself intended to capture explicitly the *phenomenon* of truth, which is not restricted to any particular language or even to contexts in which sentences are formed and uttered at all. But since it is also highly plausible, as I have argued, that Tarski's T-schema does essentially capture the formal-linguistic *aspect* of this phenomenon as it figures in the structure of natural languages, the necessary existence and implications of the semantic paradoxes within these languages must also be considered relevant here. Indeed, they might reasonably be considered to characterize in a significant sense the semantic structure of *any* language capable of discussing the truth-values of its own sentences. Thus each of the formal interpretations that make sense of the implications of the paradoxes for the overall semantic structure of a language captures at least some of the formally indicated features of the broad phenomenon of (transcendental) truth.

Certain aspects of these constructions can also be linked in more specific ways to features of the transcendental phenomenon of truth in the sense that I have described. For example, the circularity in the definition of linguistic truth to which Gupta and Belnap advert can be motivated, in the broadened context of a disclosive understanding of transcendental truth, by recalling the well-known circularity in the hermeneutic basis of the unconcealment of beings to which Heidegger points in *Being and Time*. According to Heidegger, because of the necessary involvement of the fore-structures (see above) in any interpretive understanding of the world, all possible interpretative unconcealment is itself situated within a circle, grounded in the structure of Dasein itself. In this circle, Being-in-the-world must first be understood as a whole, albeit in a vague and inexplicit way, in order that it can subsequently be made explicit.[77] In this sense, "any interpretation which is to contribute understanding, must already have understood what is to be interpreted."[78] The circular structure is, according to Heidegger, characteristic of the kind of understanding that is requisite to grasping the present-at-hand as well as other modalities of beings and is "formal-indicatively disclosed" by Dasein's understanding projection itself.[79] The circularity of the interpretive foundations of truth is thus a necessary feature of the phenomenon, according to Heidegger, and it is thus reasonable to think that any complete and explicit conceptual grasp of it would include this circularity. In particular, since assertion is, for Heidegger, a type of uncovering, itself grounded in the broader disclosive structure of Dasein, it is reasonable to suppose that when we restrict our attention to assertoric truth (as all the philosophers who follow in the tradition of Tarski do), we will find the same circular structure. Here, the fact that the circular structure of the truth-predicate for a natural language can be shown to be a consequence of the uniform applicability of the T-schema is itself an important independent confirmation of a result that could also have been predicted on phenomenological-existential grounds.

From the perspective of the Heideggerian picture, the circular structure that Gupta and Belnap discover in the concept of truth may thus be thought to reflect the fact that (due to the kind of objectification that language makes possible and its own inherent reflexivity, itself grounded in the reflexivity that characterizes Dasein) sentences of a language are, in its ongoing practice, themselves continually uncovered and considered *as* true or false. *Thereby,* they become entities that can *themselves* be discussed in terms of their *own* semantic status. In the course of inquiry, the extension of the truth-predicate will then pass through various stages of revision. At each stage, it is plausible that the status of various sentences or sentence-types as true or false is itself taken into consideration and used as a guide to the revision. This might even reasonably be thought to include moments of "infinitary" reflection involving the consideration of infinite sets of sentences of a particular type or characterizing a certain subject matter. These are the moments modeled, in Gupta and Belnap's construction, by the "limit" stages of the procession to, and through, transfinite ordinals. This provides at least a phenomenological motivation for the reality of the transfinite progression by indicating how it could indeed plausibly be reflected in the narrower dynamics of the behavior of the truth-predicate of a language over time.[80]

What, then, of Priest's second argument against the revision theory, the argument concerning the relation of the theory to the actual knowledge and practice of speakers? As we have seen in connection with Dummett, the phenomenon of transcendental truth itself does not need to be seen (and should not be seen) as necessarily grounded in the capacities, abilities, or practices of individual knowers in order to be explicated phenomenologically. It is, rather, to be seen as grounded in a structure which, though it is *formally indicated* as inherent to Dasein's facticity, also does not have to be (and generally is not) present in the implicit or explicit awareness of any individual subject or collectivity thereof. Moreover, the formal indication is itself understood as involving important aspects of formalization and interpretive articulation, so that it may turn out *on reflective interpretation* to have essential structural moments that are not at first apparent from the facts about linguistic usage or practice alone. Though the existence of elaborate formal structures such as the transfinite construction of Gupta and Belnap's picture would admittedly have to be positively demonstrated in relation to the underlying phenomenon of truth, it is thus not inconsistent with this conception that it exist and characterize, more narrowly, the meaning and behavior of the truth-predicate as it is actually employed in linguistic practice. Priest's second objection to the revision theory is thus to be rejected in this context. Similarly, to see the phenomenon of revision as actual and having a real temporal significance, it is not necessary to assume that the revision happens periodically or in a regular chronological fashion, but only that it can occur at particular moments of holistic transition, such as those in which a formerly guiding conception is reevaluated or a new holistic interpretation of a broad range of phenomena first becomes available.

Nevertheless, Priest's first objection—that the revision theory and Kripke's theory do not in fact eliminate paradox, but only relocate it—remains legitimate, and in fact *also* can be seen as pointing to significant features of the underlying phenomenon of transcendental truth. As Priest argues in detail, it is plausible that *any* theory that eliminates the semantic paradoxes and the closely related set-theoretical ones from a particular language will do so only by *limiting* the expressive resources of the language in question. It is this limitation, for example, that saves the consistency of each language within the Tarskian hierarchy by stipulating that none of the languages in question can capture their *own* notion of truth. The choice between consistency and expressive completeness is, then, a fundamental one, and it appears to be unavoidable that any theory that attempts to characterize a language as a whole must choose for one or the other.[81] Thus if we consider the language in question to have *full* expressive resources, as characterized by the *unlimited* applicability of the T-schema, we must also accept that there will always be paradoxes and contradictions that characterize the very structure of truth.

Should we, then, see truth, as illuminated by the Heideggerian picture, as structurally circular *or* as essentially contradictory? The answer is "Both." The inherent reflexivity of any language that discusses truth, itself a structural outcome of the hermeneutic circularity of Dasein, means that the semantic structure of any existing language can be viewed as containing a circular truth-predicate that is, at any moment, incomplete, and as undergoing ongoing revision as the semantic consequences of existing theories are themselves considered and reflected upon. The process is, under this aspect, never complete (even when there are noncontradictory fixed points), and will always leave some sentences in the category of paradox or instability. But it is equally possible, with a shift of perspective, to characterize the truth-predicate of a language as *already* embodying the general phenomenon of truth *as a whole*, and thereby to see the actual phenomenological basis for the necessary structural existence of contradiction and paradox. The two perspectives—that of consistency with incompleteness and that of inconsistency with completeness—cannot be occupied simultaneously; but if the structural features of truth are indeed phenomenologically characterized as I have suggested here, it must essentially be seen as describable only in terms of this irreducible duality. The inherent circularity *and* contradictoriness that thereby emerge as essential structural features of the concept of truth would be a problem, in an obvious sense, for any theory that wishes to *define* truth consistently and noncircularly. But where the goal is not the fixing of definitions but rather phenomenological and formal indication of an underlying problematic structure that is and must be presupposed, it is not unreasonable to see the structure as bearing these features essentially.[82]

In closing, it is worth noting that although the two issues I have discussed in this section and the last—namely, the issue of the status of linguistic senses as "modes of presentation" and the issue of the structure of truth-predicates in natural languages—are not the same, there is nevertheless a suggestive formal and phenomenological homology between them. In both cases, what is phenomenologically at issue is the way in which sentences reflect, in the narrowed context of linguistic assertion, the broader phenomenon of truth characterized in terms of presentation and presence. In both cases, as well, this reflection produces *irreducible* structures of circularity and contradiction within language itself. The apparent inevitability of these structures plausibly demonstrates, in both cases, the impossibility of characterizing the functioning of meaningful language wholly in syntactic terms, or of any clean divide between the syntax of language and the semantics of truth. But importantly (and as we shall see in more detail in chapters 4 and 5), this is not to say that the project of combining the two—truth-functional/quantificational syntax and ontological semantics—into a unitary, complete, and overall self-consistent picture fares any better. Rather, in light of the broadened consideration of the phenomenological conditions of assertoric truth, the paradoxical and aporeatic results (such as Tarski's) that have appeared to demand both the separation of the two and their subsequent recombination into a unified picture of meaning actually show both (the separation as well as the recombination) to be untenable, once the broader phenomenon of presentation is taken into view. The aporias and paradoxes considered here can then emerge as formal indications, within the structure of language more narrowly conceived, of the broader structure of the phenomenalization of phenomena, or the presentation of presence as such.

4

The Undecidability of Sense and the History of Being

In the last chapter, we saw how Davidson's and Heidegger's accounts might be unified to bring out a single, common structure of truth, formally indicated in the unified hermeneutic situation that allows for the intelligibility of any entity as such. This unitary structure is itself grounded, as I argued, in the ontological difference and can be formally articulated through the metalogical structure of paradox that marks the formal logic of truth in its own self-application. As we saw in the last chapter, the resulting account appears capable of indicating the grounding of the specific phenomenon of truth in both of two complementary directions: the intelligibility of entities is seen as grounded in the specific possibilities of meaning afforded by a structured language as it is spoken, but conversely, linguistic meaning is itself grounded ultimately in the holistic disclosure of entities, or what Heidegger understands as their "entry" into a world.[1] As I shall argue in this chapter, the second direction of grounding points, in turn, to an irreducibly *temporal* dynamics of sense: one that allows for languages to differ significantly from one another and themselves over historical time, differently determining the conditions of the intelligibility of entities therein.

In Heidegger's work during and after the mid-1930s, such a temporal dynamics is the basis for describing the "history of being [*Seinsgeschichte*]" as a process in which successive epochs of total intelligibility replace one another, each one determining differently the overall sense of the entities available to thought and description within it. One formal reason for this temporally dynamic structure, as I argue here, is the inherent *undecidability* that appears necessarily to characterize the total logic of a language's own characterization of its possibilities of truth. This structural undecidability arises within instituted languages, as I shall argue, at the inherent point at which these languages reflexively figure or envision their own constitutive relation to being, in the sense of truth. For whereas each such language includes within itself, as an inherent structural moment, the capacity to theorize the sense and meaning of its own sentences in general, each one also involves a going conception of these sentences as capable of truth or falsehood which relates their behavior, at least implicitly and in general, to being as such.

A consequence of this is that, in relation to new phenomena and unanticipated cases, it is never possible fully to specify unitary procedures that will always completely and exceptionlessly determine the sense of linguistic terms. But this undecidability is, on the other hand, not merely limitative, since it also points to an original ontological structure linking historically instituted languages to the structure of truth, and thereby to being itself.[2] Specifically, the undecidability of sense witnesses an essential gap, within each language or historical epoch, between the determinate procedures that decide sense (to the extent that it can be decided) within it, and its own self-conception as capable of univocal truth.

As I shall argue, this means that each such language or epoch, even as it "totalizes" beings according to determinate standards, is necessarily, and structurally, open to the possibility of its own foundational transformation. This points, further, to a temporal determination of the changing interpretations of beings that is not in competition with the antirelativist conclusions that Davidson draws from his own picture of the constraints introduced by radical interpretation. Finally, it provides support for a temporally dynamic but nevertheless realist account of the ontological foundations of sense, bearing against anthropological, culturalist, and other antirealist conceptions of these foundations.

1. The Institution of Language and the Undecidability of Sense

In the *Beiträge zur Philosophie: Vom Ereignis*, composed between 1936 and 1938, Heidegger announces a transition from the previous "guiding question [*Leitfrage*]" of philosophy, that of the *nature* of beings or entities, to the "grounding question [*Grundfrage*]" of the *truth* of being itself. The transition is announced as an integral aspect of thinking from and toward the essential happening of being, or *Ereignis*:

> The question of being [*Die Seinsfrage*] is the question of the truth of beyng.[3] When grasped and worked out historically, it becomes the grounding question [*Grundfrage*], as opposed to the previous question of philosophy, the question about beings (the guiding question [*Leitfrage*]).[4]

This new question of the "truth of beyng" is further understood as the question of an "openness for essential occurrence [*Wesung*] as such" and as the question of the ultimate ground for anything existent to be.[5] Thus thought as the ground of all possible appearance, this "essential swaying" of beyng [*Wesung des Seyns selbst*] is designated as *Ereignis* and a preliminary and transitional "thoughtful speaking" of, and from, it is attempted.[6] This attempt at thinking and speaking is, according to Heidegger, to be understood as a "directive [*Weisung*]" that indicates the "necessity" of the "sheltering of the truth of beyng within beings [*der Bergung der Wahrheit des Seyns in das Seiende*]" on the way to a possible "leap" from the previous question of beings to the new question of the truth of beyng in itself, without reference to beings.[7] Nevertheless, the task toward which it is directed remains, as in *Being and Time*, related to the specific question of the grounding of domains of beings, including that of beings as a whole. Here, for example, this task is specified as that of the "retrieval of beings out of the truth of beyng [*Wiederbringung des Seienden aus der Wahrheit des Seyns*]."[8] This task is to be carried out specifically by means of a distinctive questioning about "sense [*Sinn*]" as "the grounding of the projected domain":

> The question of "sense," i.e., according to the elucidations in *Being and Time*, the question of the grounding of the projected domain [*des Entwurfsbereichs*], or, in short, the question of the *truth of beyng*, is and remains *my* question and is my *unique* question, for at issue in it is indeed what is *most singular* [*den sie gilt ja dem Einzigsten*] . . .
>
> The question of the "sense of beyng" ["*Sinn des Seyns*"] is the question of all questions. In the development of its unfolding, the essence of what is here called "sense" determines itself, that within which the question as meditation [*Besinnung*] persists [*sich hält*], that which it opens up as a question: the openness for self-concealing, i.e. truth.[9]

The new "grounding" question of the truth of Beyng thus results from a historical and ontological deepening of the problematic of sense and truth already pursued in *Being and Time*. As we have seen (chapter 3), for Heidegger in *Being and Time*, *sense* is fundamentally Dasein's temporal projection of possibilities onto what are thereby constituted as intelligible entities. The possibility of this projection is itself closely related to that of disclosive truth, in that both take place on the ground of Dasein's fundamental hermeneutical structure of interpretation, that of the hermeneutic-existential "as." In *Beiträge*, Heidegger deepens the linked question of the basis of sense and truth into the question of the basis of projection itself, or of the kind of opening and concealing that is characteristic of Dasein as such. Dasein is itself, accordingly, no longer seen as a positive given phenomenon to be described or illuminated. Rather, it is to be achieved or accomplished by grounding *itself* in the truth of beyng. This is, in particular, to be achieved by way of a thoughtful meditation on this truth, which also has the significance of an "appropriation" of Dasein by, and into, *Ereignis* as beyng's event.

With the transition from the "guiding" to the "grounding" question, according to Heidegger, being is no longer understood in terms of, or by reference to, entities. There is then no longer any possibility for thought to decide on the being of what is, except from out of the truth of beyng. Thus, "beyng can no longer be thought on the basis of beings but must be inventively thought from itself [*muß aus ihm selbst erdacht werden*]."[10] At the same time, it is now possible to view the "metaphysical" interpretation of being hitherto as the history determined by the prevalence of the previous "guiding" question itself. Within this history as it now appears, the ongoing and increasing withdrawal of being leads to the succession of historical epochs in which beings are determined metaphysically or onto-theo-logically. This determination typically works by specifying a unifying character of "beingness," or what it is, most fully or paradigmatically, to be, as determined by a maximally universal or most elevated being, as well as a general determination of the underlying character of entities. The progressive withdrawal of being itself from the epochs determined by these specifications continues up to the point of the contemporary regime of technology or (as it is termed in the *Beiträge*) "machination [*Machenschaft*]," in which there is no longer any such single organizing referent. With this, the character of being itself which was still present, albeit always in an increasingly veiled way, in the previous organizing referents has withdrawn *completely*, leaving only the universal technical "calculability" of entities, and their immediate availability to the anonymity of "lived-experience [*Erlebnis*]" in general, as the empty forms of their general reality and sense.

The sequence of historical epochs that characterizes the history of the metaphysical interpretation of being is specified, with minor variations, in the *Beiträge* and several other texts. In the 1943 lecture "Nietzsche's Word: God Is Dead," the ontic principles that have governed collective understanding and action through the history of metaphysics, understood as the history of the basic constitution of entities as a whole [*das Grundgefüge des Seienden im Ganzen*], are given as "the supersensory World, the Ideas, God, the moral Law, the authority of Reason, Progress, the Happiness of the greatest number, Culture, Civilization . . ."[11] In the 1957 lecture "The Onto-theo-logical Constitution of Metaphysics," Heidegger specifies the "historical stampings [*geschicklichen Prägungen*]" of being within this history, somewhat differently, as "*phusis, logos, hen, idea, energeia*, substantiality, objectivity, subjectivity, will, will to power, will to will."[12] The historical succession begins with the Greek conceptions of the character of the world and thought and proceeds through Descartes's interpretation of thinking and being in terms of subject and object, eventually culminating in the configuration of contemporary technology, which Heidegger identifies as prepared by Nietzsche's metaphysics of will. Specifically, the Nietzschean principle of the will to power is here ultimately understood as a redoubled "will to will" in which entities are constantly handled, circulated, and calculated without any overarching purpose or goal except instrumentality itself.

Along with this explicit consideration of the epochal history of being, Heidegger's turn toward the "grounding question" of beyng itself in the 1930s is also marked by a renewed attention to the ontological and historical character of *language*. While, on the one hand, Heidegger now understands the epochal configurations of metaphysics as intimately related to the "historical languages" that successively articulate them, on the other he now poses the question of the "being of language" as one closely related to the radicalized "grounding" question of the truth of beyng itself.[13] During this period, Heidegger considers the investigations of Herder into the origins and expressive structure of languages.[14] These considerations can be related to a more general mystery or paradox of the origin of language in its referential, descriptive, or nominative relation to the world. In particular, prima facie there is a paradox that arises whenever language is thought to be originally connected to the world through some initial act of institution or primary naming. The paradox is that, for the requisite connections between words and things to be set up, it must apparently be presupposed that things are already understood in determinate ways. But this understanding plausibly only comes *through* language, and so *cannot* be thus presupposed. The institution of meaning thus appears to presuppose that meaning already exists, so the possible origin of significant language itself remains obscure on any such picture.[15]

For Heidegger, this paradoxical situation that arises in considering the origin of language is actually positively indicative of an ontologically characteristic feature of language in relation to its being: its tendency to withdraw or refuse itself to positive overall description.[16] Additionally, what is indicated in the paradox of origin and in the withdrawal and refusal of language to its own positive description can also be grasped as pointing toward that deeper character of language that is thought, through the Western tradition, as the character of the "logos."[17] Throughout the succession of historical determinations of the beingness of entities, *logos* names simultaneously being and saying, and thus points to the assumed basis or actual ground for the possible linguistic expressibility of things as such. Within the history of metaphysics, this determination occurs through the particular historical languages that privilege such specific determinants of the nature of entities both *as such* and *as a whole.*

How should we understand the basis of the determination whereby historical languages themselves become structured by the privileged standards or measures that in turn determine how beings "as a whole and as such" appear and can appear? In *Heidegger on Being and Acting* and *Broken Hegemonies*, Reiner Schürmann interprets Heidegger as pointing to a series of "principial economies" that have successively unfolded over the historical time. Each is stabilized by a particular "epochal principle" or "hegemonic phantasm" which functions to stabilize and orient practice by elevating a *particular* entity to the rank of an absolute measure or standard for the interpretation of *all* beings.[18] This structure of the stabilization of organizing referents replicates, at each moment of transition from one epoch to the next, the original paradox of the institution of language itself. In particular, since each epochal economy achieves, by means of the elevation of an epochal standard, the ability to refer to and express beings as a whole, each one instantiates the general structural paradox of the authority of a particular element over the totality of which it is a part. If each hegemonic phantasm "measures measures," setting up in its institution a total system of legitimacy in which beings become intelligible as such and regulating what can appear at all, then each also constitutively implies the paradox that the measure itself cannot be measured in its own terms.[19] The ultimate standard both is and is not an entity; it is the source of all measurement and regulation that, for this very reason, cannot itself be measured.[20]

If this paradox of instituting standards characterizes the "historical" being of language as such, it also characterizes the particular "historical" languages of the various epochs in their relation to beings as a whole. According to Heidegger, this relation is accomplished by means of a "projection" of sense that first renders the domain of entities as a whole accessible in some particular way. This temporal projection, in making entities accessible, prepares the possibility of determinate reference that subsequently characterizes the structure of a language overall. It thereby yields the determinate standards and principles of logic, grammar, and practice by which the speakers of a language "measure" beings in their being. Once instituted, these principles and standards provide a basis for linguistic decision on sense within the constituted domain. But they cannot ultimately decide on themselves; in particular, they cannot provide a noncontradictory basis for deciding on their *own* scope of application. There is thus a structurally necessary *undecidability* of linguistic sense, inherent to the semantical structure of each epochally instituted historical language, and which can (as we shall see in the next section) even be positively demonstrated by means of a formal analysis of the conditions of this structure.

2. Against Relativism: Conceptual Change and the History of Being

In "On the Very Idea of a Conceptual Scheme," Donald Davidson gives a well-known and far-ranging argument against the "heady and exotic" doctrine of conceptual relativism.[21] In particular, Davidson argues that it is incoherent to suppose that different systems of categories, conceptual perspectives, or conceptual organizations of experience are embodied in "conceptual schemes" that are both identical with or contained within particular languages and significantly distinct from one another. The basis of Davidson's argument is his inquiry into the structure of linguistic interpretation, which forms the methodological core of his influential program for the provision of theories of meaning for natural languages.

In this program, Davidson requires that the theory of meaning must be worked out under the condition of "radical interpretation." Here, the interpreter has no initial knowledge of the language to be interpreted and must reconstruct its meaning on the basis only of the intersubjectively available evidence, including speakers' assent or dissent to particular sentences under particular conditions. Under this constraint, as Davidson argues in "Belief and the Basis of Meaning," it will be impossible actually to interpret the language of a speaker unless a significant amount of *agreement* between the interpreter and the interpreted is actually assumed at the outset. Since it is not possible to interpret another speaker's beliefs by means of her utterances unless it can be largely assumed that they express truths, there is always a necessary trade-off of belief and meaning in interpretation. It is thus obligatory to apply what have been called "principles of charity" in interpretation: it must be assumed that the beliefs of the speakers under interpretation are largely identical with one's own in order for the interpretation even to be possible in general.[22] For this reason, the very possibility of interpretation involves a "vast amount of agreement on plain matters" and the possibility of disagreement is generally only intelligible against an assumed background of "widespread" agreement.[23] Because of the way charity thus figures in the necessary conditions of any possible interpretation, Davidson argues in "On the Very Idea," it is thus in fact impossible to judge the concepts or beliefs of the speakers of another language to be "radically" different from our own, given that interpretation is possible at all.[24]

On this basis, Davidson challenges a family of metaphors that attempt to make sense of the idea that a pattern of linguistic behavior can be seen as embodying a particular "conceptual scheme" at all. On some accounts, such a scheme is successful insofar as it succeeds in "organizing" or "fitting" the world, objects, or the "given" of experience.[25] These metaphors of organization and fit themselves have various versions, but Davidson argues that none of them are ultimately coherent. For, to begin with, the idea of a relationship of "organizing" between language and reality (or experience) presupposes the prior existence and determinate ontology of the entities that are supposed to be organized. Once this is assumed, however, there is no longer any room for a radical difference in the ontologies of different conceptual schemes. Variant conceptual schemes will then just be different ways of expressing the same ontology rather than the embodiment of radically different ones.[26] The idea of an accurate scheme or theory as being correct in that it accurately or appropriately "fits" the totality of experience or the world is similarly idle, Davidson suggests, in that we understand the appropriateness of the "fit" only as a matter of the scheme or theory being largely *true*.[27] But we understand truth only through the kind of translation or interpretation of sentences that is modeled explicitly by the corpus of Tarskian T-sentences. Once a Tarskian truth-theory for a language is given, the metaphor of "fit" adds nothing substantive. In particular, it again cannot serve as a basis for comparing what are thought of as distinct "conceptual schemes" set over against the world or the totality of experience.

Davidson suggests that the idea of a dualism of "scheme and content" embodied in the metaphors of fitting or organizing is a dogma, a "third" dogma of empiricism to be rejected along with what Quine treated as the dogmas of reductionism and of the analytic/synthetic distinction.[28] According to Davidson, this dualism of scheme and content underlies the idea that two languages or "theories" may be *incommensurable* in the sense that there is no neutral or empirical way to settle substantive differences or disagreements between them. This appears, in particular, in the example Davidson gives, to be the position of Kuhn in his interpretation of the history of science as structured by a series of successive theories or "paradigms" which replace one another discontinuously in the events that Kuhn calls "scientific revolutions." Specifically, Kuhn holds that

> in the transition from one theory to the next words change their meanings or conditions of applicability in subtle ways. Though most of the same signs are used before and after a revolution—e.g. force, mass, element, compound, cell—the way in which some of them attach to nature has somehow changed. Successive theories are thus, we say, incommensurable.[29]

Davidson suggests that Kuhn means "incommensurable," here, in the sense of "not mutually translatable": the supposed difference between two theories in their "way" of "attaching to nature" is thus taken to imply that the similar-sounding locutions used by the two theories are not in fact capable of being translated into one another without loss. But it is clear that, as Davidson now suggests, the very possibility of making sense of alternative or older theories *as* determinate theories at all undermines this claim of untranslatability. The proponents of older theories or paradigms might be seen as ignorant of phenomena of which we now know, or confused in grouping together some entities now distinguished or drawing distinctions where no real ones exist. But if we can understand their theory at all, there is no general ground for the claim of untranslatability, and Kuhn's metaphorical picture of different theories as embodying distinct "ways of attaching to nature," which again turns on the dualism of scheme and content, should itself be rejected as incoherent.

At first glance, it can seem that Davidson's argument in "On the Very Idea of a Conceptual Scheme" poses a direct challenge to Heidegger's conception of the history of being. Like Kuhn's successive "paradigms," Heidegger's successive "epochs" explicitly involve basically different global ways of understanding the nature of things and the larger contours of reality. Indeed, these differences are both farther-ranging and more deeply rooted for Heidegger than they are for Kuhn, since they not only characterize particular scientific theories or scientific practices, narrowly defined, but extend to the whole unity of possible thought and practice at any given historical time. Moreover, as we have seen, the distinct orientations or positions that are involved in each of the historical epochs, as they are described by Heidegger, are indeed conceived as both individually *total* in their determinative relationship to beings as a whole and deeply different from one another. Indeed, on Heidegger's account, each of the successive epochs or "principial economies" yields a standard or measure that, for a time, determines and regulates the appearance and relations of all entities. Since each of the different epochal principles "measures" the being of beings by onto-theo-logically grounding beings as a whole, and the measures are distinct, there is also a clear sense in which the economies of presence involved in distinct epochs are themselves "incommensurable," having no common, higher ontic "standard" that itself could be used to measure their differences.

This initial impression of conflict between Heidegger's results and Davidson's can begin to lessen, however, as soon as we consider that Heidegger has in view a primarily *temporal* configuration of diachronic change which nevertheless takes place within an essentially *unitary* ontological situation. This is the unitary situation of the onto-theo-logical history of metaphysics, determined as the history of successive interpretations of being or beingness as presence, but ultimately (as we have seen) from the truth of beyng itself. Given this unitary situation, the impression of relativism that at first may seem to be a consequence of Heidegger's picture of conceptual change can dissipate. Here, it is in fact decisive for Heidegger's own picture that every actually spoken "historical" language must be considered to coimply, as such, the world as a *totality* of beings. For this reason, it must be intertranslatable (in a broad sense) with every other such language. At the same time there are, nevertheless, different ways of making and stabilizing this constitutive reference to the totality which can shift over time while translatability is maintained.

On Heidegger's account, this kind of change may take place within languages, but it is often more characteristically indicated by shifts between them. Thus, for example, the early but distinct conceptions of *idea* and *hypokeimenon* correspond to the "Greek" configurations represented by Plato and Aristotle, each of which is obviously distinct from, but also continuous with, the "Latin" conception of a creator God as the highest being and the figure of transcendence. In an obvious sense, the organizing conception of a later epoch, or at least a sense of its "point" or "purpose," may not be directly accessible, at least not without significant indoctrination or education, to the partisans of an earlier one. In some cases, Heidegger suggests, the organizing referent of an age may even be largely invented or constructed by means of semantic shifts or new interpretations performed by philosophers, as (for example) when Plato creates a new sense of *idea* by shifting from the established sense of the "outward look" of a thing to a new sense involving the supersensible and unchanging, or when Descartes gives a fundamentally new significance to the *cogito* as thinking substance.[30]

Although these shifts can mark profound changes in the interpretation of beings, determinative for action and practice, there is no evident reason to suppose it actually impossible to *understand* any one of the total configurations from the perspective of another that succeeds it. Rather, since interpretation is guided in general not only by the specific epochal configurations themselves, but broadly and more deeply by the unitary character of being in its opening and projection of them, it is apparently possible to understand both the epochs themselves and their succession in a unitary way. Because of the way the epochs or principial economies thus succeed each other temporally within the unitary configuration of metaphysics, there is accordingly no reason to suppose that this involves any actual *untranslatability* between them. The idea of incommensurability, in the sense of nontranslatability, which Davidson attributes (whether correctly or incorrectly) to Kuhn, thus does not appear to figure in either the motivation or ultimate implications of Heidegger's picture. Nor is there, on Heidegger's picture, any dualism of scheme and content, with respect either to particular languages or to language in general. Rather, as we have seen, a language is integrated into the world as a single, holistic system of meaning, grounded and exhibited in the specific possibility of interpretation, that simultaneously gives meaning to words *and* to the things they describe.

This holistic picture of the integration of language and world is in fact familiar from Davidson's own conception of linguistic behavior as intrinsically integrated with practice and with worldly objects and phenomena. For this inherently world-involving structure as Heidegger describes it, "syntactic" components cannot simply be separated from "semantic" ones, and the structure of a language as such, is, as we have seen, constituted by means of the projective understanding that renders entities intelligible according to a particular interpretation of their being.

What, then, of the important discontinuities between different epochs of presence? Here it is again important to note that what is at issue in Heidegger's picture is not simply the meeting of languages or schemes considered different from each other and juxtaposed, but also the *internal* development of an essentially *continuous* trajectory. In the course of this development, older epochal principles and the economies they organize are not simply rendered inaccessible or the conceptions they have organized incomprehensible from the perspective of newer ones. Rather, they become specifically visible, in their destitution, as the ontic referents they always in fact were.[31] At the moments of transition from one principial economy to the next, this makes for a specific dynamics of transformation which, as we have seen, becomes fully visible only retrospectively, from the position of the new understanding looking backward. But the structure of such transitions is itself, on Heidegger's account, recurrently determined by the underlying structure of being in its granting and withholding of presence. This granting and withholding determines, in each case, the total measure of beings that a particular epochal principle allows and maintains.

By noting these essential features of stabilizing referents and the theoretical structures they produce, it is possible to see in them formal grounds for an alternative to standard accounts of the dynamics of "theory change." These accounts often refer such change to political or sociological factors, themselves conceived as simply external to rational and empirical scientific inquiry.[32] By contrast, the dynamics of large-scale change we have considered can be characterized both formally and "ontologically": since, as I have argued, they are constitutively related to the underlying presentational structure of sense itself, these dynamics are plausibly involved in any historically constituted linguistic situation, and not just those in which certain contingent sociological or political features are present. But the specification of the general dynamics as formal and ontological in this sense does not, at the same time, exclude their constitutive involvement with the actual phenomena or processes discovered in the course of empirical inquiry. The picture is, rather, that such inquiry, as an ongoing process that always also involves ongoing conceptual reflection on its results, includes the complex conditions of truth as disclosure, up to and including the kind of transformative disclosure in which the overall character of entities changes in a basic sense.

On the suggested picture, positive inquiry into any "domain" is impossible and basically unintelligible if this domain is not first *opened* to the possibility of inquiry by means of an ontologically based projection of sense. Nevertheless, this projection is not simply "prior" to empirical inquiry and discovery; rather, it is an inherent structural moment of the larger holistic phenomenon of interpretive disclosure, of which ordinary empirical inquiry is *also* an integral part. In this way, as I have argued, it is possible to see even revolutionary changes in paradigms and theories as determined *structurally* from the implicit dynamics of instituted sense, insofar as they are formally and ontologically related to truth.

In fact, far from being directly opposed to it, the suggestion of a temporal and paradoxical dynamic of sense, insofar as it is related to truth, receives motivation from another structural part of Davidson's own Tarskian account of truth and meaning. As we have seen, Davidson follows Tarski in holding truth and meaning to be systematically interconnected in the structure of a language. Specifically, for Davidson, under the condition of radical interpretation a theory that gives a true Tarskian T-sentence for each sentence of the language will clarify the structure of meaning for the language as a whole. The application of such a structure to any natural language will also involve, however, that some of the sentences whose meaning is thus characterized themselves involve reference to language, and in particular to truth. And as is well known, the inclusion of the possibility of discussing truth within a language leads directly, when combined with the general apparatus of the T-sentences themselves, to structural paradoxes.

The most basic of these paradoxes is that of the Liar, the sentence which says of itself that it is not true. But as we have seen (chapter 3 above), more broadly, a consequence of the inclusion of a truth-predicate within a natural language (which is also capable of forming names for its own expressions) is that the totality of T-sentences for such a language is not consistently decidable: that is, there is no finite procedure capable of consistently determining the totality of the (true) T-sentences. This undecidability remains, as we saw there, *even if* explicitly "paradoxical" sentences such as the Liar are treated as meaningless and so as neither true nor false (for instance in connection with an intuitionist or antirealist semantics of them). For even then, it must be possible to construct sentences involving the criterion of meaningfulness itself, which will again render the T-sentences as a whole undecidable.[33] In the context of Davidsonian assumptions about the relationship between the structure of the T-sentences and natural-linguistic meaning, this points to a basic phenomenon of the *undecidability* of linguistic sense, whereby meaning in the sense of a Tarskian truth-theory for *any* language that includes its own truth-predicate and the general ability to make reference to its own sentences will fail to be consistently decidable by means of any finite procedure.

On the usual model-theoretic interpretation of Tarski's construction of theories of truth for formal languages, such a theory is grounded by first specifying relations of "satisfaction" for primitive terms and predicates. Within the Tarskian construction, the satisfaction relations function as axioms for the recursive definition of the truth-predicate by specifying formally in detail what is intuitively the "reference" of simple nominative terms and the extensions of primitive predicates. It is crucial to the construction that there be only finitely many such axioms, considered as giving the extensional definitions of simple or "primitive" terms of the language's (finite) vocabulary; otherwise there will be no tractable explanation for the infinite capacity of the language to produce new sentences capable of truth or falsity.[34] The axiomatic specification of these satisfaction relations might be thought to be at least analogous to the moment of the institution of a language in its specific relation to the world; through such a specification, the language is specifically defined in its "semantic" bearing on reality. But the structural paradox introduced by the inclusion within a language of its own truth-predicate means that no language thus instituted can ultimately be axiomatized in such a way as to render its sense decidable.[35]

It appears to follow that the sense of the terms and sentences of such a language, even if conceived as a direct outcome of the organizing principles and epistemic procedures constitutive of the language as such, must itself ultimately be *undecidable* in terms of these very principles and procedures. The constitution of particular languages in terms of ultimate principles that function axiomatically cannot, then, ultimately stabilize meaning. The further question of the ground of the principles that govern sense points to a deeper phenomenon of truth that then cannot be completely captured by any consistent set of intralinguistic principles, standards, or procedures. In connection with such principles and procedures, there is thus always a residual undecidability that becomes explicit in the question of *their* grounding and points to the always-open possibility of their radical transformation.[36] In this way, the specific axiomatic or principial constitution of languages, which first makes them capable of objective reference to a totality of entities and stabilizes the ontological sense of this reference by providing a ground for decisions on meaning, always also evinces the structural undecidability of sense. This undecidability then points to the deeper conditioning of all standards and (ontic) grounds in the more original structural context of irreducibly *ontic-ontological* truth.[37]

3. Historical Temporality of Language and World

As I have argued, the kind of difference that characterizes the transitions between the successive epochs of metaphysics for Heidegger is not the "conceptual relativity" that Davidson opposes in "On the Very Idea." Once we grasp the kind of diachronic unity that is, on Heidegger's account, at the structural basis of the conditioning of epochal truth by the deeper underlying phenomenon of the truth of being itself, it is clear that his conception avoids relativism. It does so by basing itself on the inherently paradoxical dynamics of possible change, which themselves result directly from the way that *every* metaphysical language grounds beings as such and *as a whole*.

In this way, the Heideggerian account also provides the elements for an account of the *temporality* of linguistic change that is essentially absent from Davidson's picture. It is not that Davidson ever actually denies that languages originate at a particular time and are transformed in various ways over the course of their careers; he just does not address the issue. What form, then, can an account of linguistic transformation take, if it acknowledges the paradoxical ontological dynamics of totality and undecidability discussed here? In the 1928 treatise "On the Essence of Ground," Heidegger considers the way in which the availability of entities for positive reference and description is grounded, in each case, in the phenomenon of projection whereby Dasein opens a particular domain through a projective understanding of the being of the beings.[38] The ultimate condition for this opening is to be found in the structural "transcendence" of Dasein, whereby it is always already "outside itself" in its original structural relation to truth as unconcealment and thereby *grounds* in its own structure the ontological difference between beings and being.[39] In this sense, according to Heidegger, the availability of any domain of entities is always conditioned by a prior interpretation that makes accessible a particular domain on the basis of an interpretation which is itself grounded in Dasein's (initially inexplicit) understanding of being. In this interpretation, new entities may "enter" the world through sudden shifts that allow something of the character of a domain of beings, hitherto obscure, to appear. The phenomenon of "world-entry" is thus to be characterized on ontological grounds as a basic possibility of Dasein insofar as Dasein itself is "world-forming [*Weltbildend*]," that is, structured by a basic transcendence that relates it to the ontological difference.[40]

But what is it to form a world? In the 1930s, Heidegger replaces the earlier conception of truth as grounded in the unitary structure of Dasein's "transcendence" with the diachronic conception of a plurality of successive historical epochs. But the succession of epochs remains, as we have seen, grounded *as a whole* in the unitary configuration of the onto-theo-logical interpretation of being as presence and the progressive withdrawal of being itself within the ambit of this interpretation. In 1935, in "The Origin of the Work of Art," Heidegger develops, on the basis of a consideration of the temporal and ontological nature of artworks, a penetrating new conception of the ontological and temporal conditions under which it is possible for "a world" to be "set up" or "opened up" through the "setting-to-work" of a work of art. On this conception, the process of "setting up" by which worlds are formed is a "worlding" of the world whereby a whole domain of objects, practices, and possible perceptions is first made available.[41] This possibility of "opening up" a world is itself understood as arising from the essential "opposition [*Gegeneinander*]" that Heidegger describes as the "strife [*Streit*]" of earth and world.[42] This strife is, on Heidegger's description, the tension or rift [*Risse*] that makes it possible for a work not only to "set up a world" but to "set forth the earth"; it is in this way that "historical man [*geschichltliche Mensch*]" grounds life in the world.[43] The grounding is, specifically, an "instigating of the" strife and thereby a "setting to work of truth."[44] Truth is, here, understood as the "essence" of the true, and in particular as *aletheia* and unconcealment.[45] But that there is truth as unconcealment means that there is "already manifest something to which we can conform ourselves."[46] The more original condition for this manifestness is what Heidegger calls the clearing [*Lichtung*]:

> And yet—beyond beings [*über das Seiende hinaus*], not away from them but before them, there is still something else that happens [*geschieht noch ein Anderes*]. In the midst of beings as a whole an open place occurs [*Inmitten des Seienden im Ganzen west eine offene Stelle*]. There is a clearing [*Eine Lichtung ist*]. Thought of in reference to beings [*vom Seienden her gedacht*], this clearing is more in being than are beings [*seiender als*

> *das Seiende*]. This open center is therefore not surrounded by beings; rather, the clearing center itself encircles all that is [*alles Seiende*], as does the nothing, which we scarcely know.[47]

The "setting up" of a world that a work can accomplish is thus related back to the specific phenomenon of the strife between earth and world and thereby also to the deeper condition of the clearing, which itself "opens" all that can be discussed or considered as existing. This opening is the constitution of world from and in the dynamics of opening and setting up that the work exemplifies in its "setting to work" of truth.

As commentators have noted, the late Heidegger's grand picture of Western history as constituted and exhausted by the series of epochal transformations, ultimately "sent" or "granted" from or by being itself, stands in some tension with culturalist pictures. On these pictures, by contrast, it is essentially *contingent* human cultural practices or institutions that set up the conditions under which entities are understood and experienced in particular and differing ways over historical time.[48] On the other hand, Heidegger himself may seem at times to suggest at least aspects of such a culturalist picture of the origin and structure of worlds. For example, his own description of the specific phenomenon of world as constitutively related, in each case, to a "historical people" may seem to suggest that he understands "worlds," in the plural, in each case as an instituted correlate of localized practices or of the languages that surround those practices. On this kind of picture, the availability and intelligibility of a "whole" of beings themselves is first constituted by such a contingent configuration of practice and language. Further, the "setting up" of a world that first "discloses" such a totality is to be understood as primarily a matter of the conventional or cultural institution of practices, as such in principle comprehensible in sociological or pragmatic terms.

It is not that commentators who hold this kind of picture simply deny that the later Heidegger also characterizes the conditions for the institution of world that relate it ultimately to the "destining [*Schicken*]" of being itself and thus to ontic-ontological conditions that, on Heidegger's own account, explicitly cannot be rooted exclusively in the phenomena of "culture" or cultural "practices."[49] But it has nevertheless been tempting, given the apparent obscurity or mystical character of the notion of "Being itself" and its "destining" of history, to correct Heidegger by supplementing or replacing his conception of world-formation with some version of such a culturalist or pragmatist picture. Here what is seen as most obscure is exactly *how* something like "Being itself" could be responsible for those large-scale shifts in which the entire character of the world appears discontinuously to change, or for the evident differences between such large-scale understandings that appear to occur over historical time.

By contrast with this, however, I have argued that the institution as well as the destitution of languages and practices themselves can and must be seen, in an ontological context, as subject to an inherent structural necessity that also has its ontological and temporal ground in the structure of disclosive truth. Here, far from being obfuscatory or mystifying, a consideration of the determining ontological dynamic of truth as unconcealment is in fact essential. This dynamic does not have to be explained in mystifying, mystical, or nostalgic metaphors in order to be clarified in its underlying structure. To the contrary, it can, as I have argued, be put on a firm metalogical basis by considering the specific implications and dynamical paradoxes of the institution of languages by means of the fixation of epochal referents. This points to an underlying determination of the historical constellations in which beings appear that is not simply culturalist or pragmatic, since it itself results from the ontological and hermeneutic structure of disclosive truth that is plausibly also the underlying ground for any possible formation of cultures and practices to begin with.

4. Formal-Ontological Consequences: Inclosure and Contradiction

I have argued that the temporal dynamics of epochal change within the history of metaphysics can best be understood by means of the formal structure of paradox that characterizes *each* epochal economy in its institution of hegemonic standards. But how can we then understand, in similar terms, the broader formal-ontological situation determining this history itself, *as a whole*? As I shall argue here, doing so requires keeping in view not only the conditioning of specific epochal economies through the elevation of ontic referents, but also the paradoxical structure of ontological sense itself, through which (in Heideggerian terms) being grants itself as *Ereignis*. In relation to the successive determination of principial economies as well as the totality of the history of metaphysics, this paradoxical structure can be understood specifically as that of what Graham Priest has termed an *inclosure* paradox. Such a paradox is one at the limits of totality, arising from the relationship of a totality to a reflexive element that can thereby be shown to exist both within and without it. Moreover, by viewing the determination of the history of metaphysics in this way, I shall argue, we can clarify and specify the form of its specific limit, and thereby begin to indicate its specific "outside." It here becomes possible to begin formally to contemplate the structure of *Ereignis* as it may appear outside or beyond the specific history determined by the succession of principial economies, up to the contemporary regime of technology. One specific consequence of this is that it becomes possible to trace the ontological *limits* of the unproblematic applicability of the (logical or metaphysical) *principle of noncontradiction,* which has long been assumed and held in force (though in various nonequivalent forms) as a maximally fundamental regulative or "normative" basis for logical and rational thought, or for the possible being of entities as such.

With the ontological inquiry, it is thus possible, as I shall argue, to expose the principle of noncontradiction to motivated grounds of its immanent critique. The aim of this critique is not to depose the principle or even to challenge (for instance in irrationalist or antirationalist fashion) its force or authority over thinking, language, or reason. It is, rather, only to clarify ontologically the basis, character, and ultimate scope of this force and authority.[50] One source of contemporary questioning of the law of noncontradiction is recent logical work that has demonstrated the possibility and coherence of *paraconsistent* logics, or logical calculi in which contradictions can be stated without involving (as they do in classical logic) the (presumably unpalatable) consequence that every claim follows from them. The availability of these logics provides support for *dialetheism,* or the position that there are at least some contradictions—claims of the form A ^ ~A—that are true. In *Doubt Truth to Be a Liar,* Priest examines, from such a perspective, the arguments that Aristotle gives for the law in book 4 of the *Metaphysics* after stating it and describing it as the "most certain of all principles."[51] Priest finds that these arguments are mostly circular and that most can be accepted by a dialetheist anyway. Relatedly, Aristotle's formulations of the law in the *Metaphysics* already appear systematically to equivocate between ontological, logical, and psychological prohibitions (whereby contradictory *objects* or *states of affairs,* contradictory *propositions* or *truths,* or contradictory *beliefs* are, respectively and variously, prohibited or proscribed), and this ambiguity appears to persist in contemporary formulations.[52]

I emphasize that the critical aim of the present investigation is not to argue for or defend dialetheism in general or to argue that any specific logical system—including paraconsistent ones—is globally right or correct. It is also not to argue for the general replacement of "classical" negation with any of the alternative structures that can be defined in light of contemporary logics. Nevertheless, I take it that a critical investigation into the basis for the force of the principle of noncontradiction is possible on ontological grounds, and that the availability of paraconsistent logics shows, at any rate, that it is not simply incoherent to talk about contexts and situations whose structural logic is significantly characterized by contradictions at the limits. Gaining clarity in this investigation requires consideration of a specific conception of *negation* that becomes dominant early on and remains decisive for the "logical" tradition: that of (what I shall call) *topo-logical* negation.[53] A topo-logical picture of negation, as I shall use the term here, is any that sees negation as the drawing of a delimitative line of demarcation between two regions within an *already determinate and complete* space of concepts or contents.[54] It is as posing such a picture that we may understand, for instance, Spinoza's *omnis determinatio est negatio* (all determination is negation) as well as Hegel's use of this principle to produce the idea of a dialectic of *determining* negation.[55] But it is equally apparent in Wittgenstein's conception of negation in the *Tractatus,* according to which the totality of "positive" and "negative" facts jointly articulate reality and the world as a whole within "logical space," and a contradictory proposition says nothing since it "admits [*zulässt*]" no possible situation.[56] With the topo-logical picture of negation, the negation of the attribution of a property to a thing or of a "positive" proposition is pictured as complementary and exterior to it, an indication of what is *other to it* or *different from it* in general within the totality of a coherent and enclosed (finite or infinite) space of concepts or contents.[57] Here, then, a genuinely contradictory proposition is impossible, since there is no way for it to have a positive indicative, descriptive, or referential value with respect to anything in this space.[58]

This topo-logical picture of negation has a historical and conceptual origin in the Eleatic Visitor's solution to Parmenides's problem of nonbeing in the *Sophist.*[59] According to this solution, to say that something is non-X is to mark its difference from everything that is X, and nonbeing itself is assertible or predicable only on the basis of the superior mixing of the "great type" of difference with being and the other great types. The formal or ontological basis of negation is thus the differentiation of contents within a logical space whose internal consistency and correspondence to the totality of objects and phenomena is presumed in advance. The conception is subsequently consolidated and entrenched in Aristotle's conception of logic and categorical meaning, according to which the logical structure of judgment is one of *synthesis* and *diaeresis*, or the combination and separation of preexisting meaningful elements, and the total space of possible judgments is structured in advance by means of the total system of categories or genera, and their specific differences.[60] In picturing this space of contents as in itself total, consistent, and correspondent to the meaningful possibilities of things, the topo-logical picture presupposes it as a field of coherent and ultimately consistent presence in general.

By contrast with this, within the history of metaphysics as Heidegger describes it, being recurrently "grants" or grounds the sense of beings, as such and as a whole, through the paradoxical institution of epochal economies. This granting or grounding has the form of a provision of ontological sense for what is, again, the totality of beings. But it is itself structurally paradoxical in that the position from which sense is ultimately granted cannot be measured without contradiction, or rendered decidable, from within any of the particular economies themselves.

This position—that from which being, in the history of metaphysics, recurrently "grants" itself as the "event" of its truth—is thus not one which can fully be understood within the ambit of the unproblematic application of the law of noncontradiction. How, then, *can* we positively understand this position both logically and ontologically? It is in terms of this question that we can see, at least in one of its aspects, the ontological significance of the most notorious problem that historically arose for, and vitiated, Frege's logicist project: the problem posed by the paradox discovered by Russell in 1901.

As is well known, Frege's attempt to ground arithmetic on logic and naive set theory in the *Grundgesetze der Arithmetik* involved his formulation of five axioms or basic laws, the fifth of which—Basic Law V—required (when restricted to concepts) that two concepts, F and G, have the same extension just in case they apply to exactly the same objects. In the context of related assumptions also central to Frege's project—including the assumption that every function must be defined for *all* objects and the assumption of a *universal* comprehension principle, whereby every linguistically defined concept picks out an extension—this produces the antinomy of (in set-theoretical language) the set of all sets that are not self-membered; such a set is a member of itself if it is not, and is not if it is.

Frege's conception of logical contents requires the intelligibility of a *total* domain of thoughts or sentential *senses*, a total domain of *referents* whose properties and relations determine the truth and falsity of these senses, and a correspondence between the two domains such that no sense is both true and false and no object bears contradictory properties or relations. That there must be such a correspondence between these totalities does not mean, of course, that sentential senses *individually* must uniquely represent or correspond to particular objects or configurations thereof—Frege has, as we have seen (chapter 1 above), rejected any such assumption on the basis of his more general antirepresentationalism—but only that the properties and relations of objects must be codeterminable with the truth and falsity of sentential senses in such a way as never to produce a contradiction. But it is the assumption of such a noncontradictory correspondence between the two totalities that Russell's paradox and the related set-theoretical and semantic paradoxes most directly undermine. In particular, given the paradox, it is no longer possible to suppose that the coherent sentences whose truth or falsity must be intelligible—including eminently those involving "impredicative" terms such as "language," "truth," "sentence," or "proposition" and the reflexivity they exhibit—can always be determined as noncontradictorily true *or* false on the basis of the referents they involve.[61] This is because, as soon as the relevant concepts are considered as having determinate extensions at all, they cannot be insulated from the paradoxical effects of their impredicativity, and devices of prohibition or regimentation (such as type theories) do not resolve the problem but only prevent it from appearing.[62]

Metalogically speaking, this situation can be seen as witnessing an inherent structural *disharmony* of intension with extension. The general assumption of the coherence of impredicative concepts leads inherently to the excess by which, given the determinate extensions which this coherence implies, there is always already something that is inconsistently, both within and without them.[63] Seen this way, impredicative concepts bear witness, more generally, to the *inherently paradoxical* situation of thought in reference to a totality that must include itself, insofar as thought (or language, or the formation of sentences, etc.) takes place within the world at all.

In *Beyond the Limits of Thought* (2002), Graham Priest has argued that Russell's paradox exemplifies a general structure that recurs in a wide variety of philosophical considerations of the limits of thought, language, or reality. This is the structure of what Priest calls an *inclosure* paradox, suggesting that such paradoxes are virtually ubiquitous whenever the boundaries of totalities, such as the totality of what can be thought or said, are considered reflexively from a perspective within those boundaries themselves. Specifically, an inclosure paradox is generated, according to Priest, whenever three features are satisfied. The first is "Existence": a totality of some sort (e.g., the totality of what can be thought, or what can be said, or what is) exists. The second is "Transcendence": there is some "diagonalizing" operation which, given an arbitrary subset of the totality (or an arbitrary one satisfying some property also satisfied by the totality itself), generates another element that is *not* inside that subset. The third is "Closure": the element generated by the operation nevertheless *is* inside the relevant totality itself. Given these conditions, it suffices to apply the relevant "diagonalizing" operation to the totality itself: what will be produced is an element that, paradoxically, both is (because of Closure) and is not (because of Transcendence) within the relevant totality. In the case of Russell's paradox, the relevant totality is the totality of all sets, V;

the "diagonalizing" operation generates, given a set X, the set of all its members that are not self-membered; and Closure is assured since this is itself (in each case) a set. Then, we need only apply the operation to V to obtain Russell's paradox.

As Priest notes, a frequent response to this kind of paradox is simply to deny the existence of the relevant totality: to claim, for instance, that there is no such thing as the totality of all sets, or the totality of all that can be said or thought, or that there is no world as a totality of all that is. Proponents of this response have often pursued it along with the strategy that Priest calls "parameterization." In the face of the apparent contradiction concerning the relevant totality, the strategy is to argue that the application of the diagonalizer to it will produce a new totality that is more inclusive or at a higher level. This process (so the parameterizer holds) can be continued potentially *in infinitum*, but will never produce an *absolute* totality of the sort that would generate limit-contradiction through the self-application of the diagonalizer.[64] This is the strategy used by Kant himself in response to the cosmological antinomies, for example: in their face, Kant denies that the world, as a totality of phenomena, exists as a whole, while also maintaining that phenomena, as appearances, have a deeper basis in the (unknowable) realm of things-in-themselves. Similarly, as Priest shows, the strategy of parameterization has been employed by standard formulations of set theory, which hold that the set-theoretical totality V does not exist or is not a set, and by almost all extant responses to the "semantic" paradoxes, including the Liar paradox (here, parameterization is exemplified by Tarski's hierarchy: the strategy is to deny the existence of a universal or general definition for truth, holding that truth-in-L can only be defined in a more inclusive language L′, and so forth).[65] In this way, in each case, consistency is preserved, but at the significant cost of the denial of the existence of the relevant totality (at least as a "completed" whole).

As Priest argues, parameterization and the denial of these totalities are highly implausible, in these and other cases, for various reasons. Beyond the desire to avoid structural contradiction in the characterization of the functioning of its semantic predicates, there is no good reason to think, for instance, that a natural language, which apparently bears fully the resources to talk about itself as well as to formulate any relevant reasoning about its own semantic concepts, must instead be understood as actually comprising an open hierarchy of distinct languages, or relying on such a superior hierarchy in order for it and its notion of truth to be discussed.[66] Quite to the contrary, as witnessed by the apparent inevitability within natural languages of locutions involving "thought," "language," and the like, these languages appear to include essentially the capacity to refer to the totalities whose conceptual coherence would be denied by the parameterizer. More generally, Priest argues for the plausibility of the general principle that he calls the Domain Principle. According to the Domain Principle, any discussion of a purportedly "open" or *potential* infinity—such as the parameterizer wishes to deploy—in fact presupposes, for its own intelligibility, the prior existence of a correspondent *actual* infinity as a "completed" totality. Thus, "totalisation is conceptually unavoidable."[67] For even the parameterizer's *denial* of the existence of the absolute totality presupposes, for its own meaningfulness, the conceptual coherence of (talking about) that totality itself. Priest argues that Cantor himself accepted the Domain Principle with respect to set theory, and suggests that Kant also invokes it in some places, despite his employment of parameterization in the case of the antinomies.

We can begin to see the ontological implications of the inclosure structure by considering the application of this line of thought to the consideration of the relationship of beings or entities to their (ontologically determined) *sense.* This time, the relevant variable term is "a being" or "something that is." In previous chapters, we have explored the idea that the sense of beings—their availability not only to language and rational consideration but to any kind of holistic significance—can be understood in terms of the phenomenological and ontological status of their overall way of being presented or given.[68] In Heidegger's jargon, this way is sometimes called their "being" or their "ontological sense [*Seinsinn*]." With this terminology in mind, it is possible to pose the question of the *being of beings*: that is, the question of the ontological sense of beings as a whole. In other terms, this is just to pose the question of *how* the total domain of beings as such is given. It is easy to see how the question of the being of beings, on this formulation, already includes the basic structure of the inclosure schema. The relevant totality is the totality of beings, or whatever *is*: the relevant "diagonalizer" is "being of . . ." Owing to the ontological difference, the being of beings is not, itself, a being (Transcendence). But (as Priest himself points out in his discussion of Heidegger), it can be discussed, and thus apparently is (in some sense yet to be clarified) an object of reference, or a being (Closure).[69] This is sufficient to produce the inclosure contradiction: that being both is, and is not, a being.

It is also evident that the strategy of parameterization, in application to this question, would be idle. For it would involve denying that the reference to the totality of beings—of whatever is—is so much as *coherent.* But this reference is of course characteristic of philosophical inquiries into ontology from Parmenides onward. What is more, in view of the Domain Principle, to pursue parameterization here would involve denying the existence of a basis on which the variable term here—"something that is" or "a being"—itself has a determinate general sense. It would thereby also involve denying determinate and general sense to expressions such as the "is" of existence and the "is" of predication. To pursue the strategy of parameterization here, in other words, would essentially be to foreclose the question of the sense of being, to the reopening of which Heidegger's effort in *Being and Time* is primarily directed. If, on the other hand, the structure of inclosure along with its attendant structure of limit-paradox is here seen as ontologically indicative in a basic way, it is possible to see it as indicating a basic structural feature of the difference between being and beings—the ontological difference—itself.

The assumptions involved here could certainly be denied. Perhaps "being" does not have any unitary sense across all ontic domains. Or perhaps (contrary to the conclusions of both Heidegger's and Davidson's arguments about truth and predication, as we saw above), it does not have a unitary sense that links the issues of predication, truth, and existence in any important way. Or again, even granting "being" a unitary sense across ontic domains and contexts, perhaps there is no significant *ontological* question to be raised about the basis of this sense. Perhaps, that is, there is just no way to clarify *how* this sense is given and how it structures the *totality* of entities that it characterizes.[70] But to draw any of these conclusions is to deny the intelligibility of Heidegger's question of being itself. It is, in other terms, simply to refuse to pursue the question of the *sense* of being, such that this sense determines the being of beings as such and as a whole.[71]

If this pursuit is carried out, on the other hand, the topo-logical picture of negation is apparently overcome, at least with respect to the ontological consideration of the sense of beings as a whole. For to construe the structure of Russell's paradox and other inclosure paradoxes positively is to interpret negation, in the sense relevant to it (the sense in which the Russell set is the set of all sets *not* members of themselves), otherwise than as a distinction *within* an already constituted totality of concepts or contents. To construe the inclosure paradox positively is rather to introduce a wholly *new* "content," itself delicately situated at the limits of the relevant totality, and indicative of the totality's paradoxical form. In a similar fashion, to interpret the ontological difference as I have suggested here is to consider its own essential internal negation—that being is *not* a being—otherwise than as a demarcation between mutually exterior regions within an already-constituted field of contents. It is rather, as I have argued, to understand the ontological difference as itself an original and positive source for being's recurrent projective granting of the sense (and truth) of beings.

This implication is relevant not only to the local question of the ontological significance of inclosure paradoxes but also to the broader question of the force of the law of noncontradiction. The argument here is this. The law of noncontradiction formulates a *general* prohibition of contradiction: according to it, contradiction (in any of a variety of senses) is *as such* impossible, unintelligible, or incapable of being formulated, asserted, or understood (etc.).[72] However, in application to the kind of inclosure structure considered above—which is, as I have argued, indicative of the structure of the ontological difference itself—such a general prohibition of contradiction involves the limitation of the Domain Principle and of (naive) comprehension itself: it must be held to be impossible (for some motivated reason) to specify the ontological basis of beings as a whole, or their sense. For similar reasons, it must also apparently be impossible, then, to discuss linguistic *senses* themselves in general or as such.[73] The only way to avoid either conclusion is to deny, by means of some variety of parameterization, the existence of such a basis and to attempt to limit comprehension by means of the denial. But this limitation and denial are the same, as I have just argued, as the denial or foreclosure of the ontological question.

The general maintenance of a domain in which the law of noncontradiction can comprehensively apply must then involve the denial of the (full or general) sense of the ontological question of being itself. But this denial is (in Heidegger's jargon) none other than the "forgetting" or withdrawal of the question of being that is, according to Heidegger, characteristic of the "metaphysical tradition" since Plato. If, on the contrary, the development of the significance of inclosure as basically indicating the underlying structure of ontological difference is pursued and followed out, it thereby becomes possible to see one of the ways in which this tradition is itself historically constituted. It is constituted, in part, by means of the foreclosure and repression of ontological difference that proceeds by way of the general affirmation of the principle of noncontradiction.

Part 2

Time

5

Metaformal Realism and the Ontological Problematic

1. A Realism of the Real

In his 1951 Gibbs lecture, "Some Basic Theorems on the Foundations of Mathematics and Their Philosophical Implications," drawing out some of the "philosophical consequences" of his two incompleteness theorems and related results, Kurt Gödel outlines a disjunctive alternative which, as I shall argue here, captures in a precise way the most significant consequences of contemporary formal/ontological reflection in considering the relationship of formalism to being:

> Either mathematics is incompletable in [the] sense that its evident axioms can never be comprised in a finite rule, i.e. to say the human mind (even within the realm of pure mathematics) infinitely surpasses the powers of any finite machine, or else there exist absolutely unsolvable Diophantine problems of the type specified (where the case that both terms of the disjunction are true is not excluded so that there are, strictly speaking, three alternatives).[1]

As I shall argue here, Gödel's argument in the Gibbs lecture, if assayed in its broader ontological implications, can be understood as embodying (what I shall call) a *metaformal* reflection on the relationship of the *forms* of thought to the *real* of being in itself.[2] If it is understood in this way, the argument motivates a particular kind of *realism*—what I shall call a "metaformal" realism—which can support, in the context of the ontological problematic of being and beings, an *ontological* realism about being and (as a consequence) a *temporal* realism about given time. This is, as I shall argue, directly opposed to each of various varieties of antirealism (whether motivated by idealist, subjectivist, humanist, conventionalist, or culturalist considerations) about the topics of being, time, and their relationship.

The type of realism I shall defend here is not a realism *about* any particular *class* or *type* of objects or entities. Neither is it, primarily at least, a realism about entities as a whole. Thus it is not, *a fortiori*, an *empirical* or "ontic" realism. Nor is it a naturalism, a materialism, or a physicalism, although I also do not think it is *inconsistent* with, or even in any deep tension with, the positions that march under these banners. Epistemologically and phenomenologically, its primary motivation is not any empirical fact of contingent experience but rather what might be called the experience of *formalization*, both insofar as this experience points to the real (but contradictory, as I shall argue) point of the *actual relation* of thinkable forms to beings as a whole, and insofar as it schematizes, in results such as Gödel's, the intrinsic capacity of formalization problematically to indicate its *own* constitutive limits.

As such a position on the relationship of thought to being, it is (as I shall argue here) relevant to Heidegger's ontological problematic, and even requisite for a furtherance of this problematic on realist terms today. I take no position here on the exegetical claim that Heidegger, after *Being and Time,* increasingly seeks to distance himself from and repudiate a residual "anthropologism" or humanism, tending toward antirealism, that may still be suggested or implied in the "preparatory" fundamental analysis of Dasein there.[3] But whether or not one agrees with this claim, it is clear that the sort of realism that I argue for here is at least a *possible* position relevant to the questions of the sense of the being of beings and of the "truth" of being, independently of any relation to entities. It also provides a concrete formal basis for critical arguments and positions that are unmistakably Heidegger's own. For on one hand, as I shall argue, the attitude or position of metaformal realism as I shall develop it here provides a formal basis for the critique of any position that puts the *representing subject* at the basis of the possible thought of being. It does so by indicating the formal-ontological configuration that first underlies the ontological possibility of there being anything like a subject to begin with. On the other hand, and on the same realist terms, it provides a concrete basis for the critique of the identification of being with effective *actuality* [*Wirklichkeit*], which Heidegger sees as characteristic of the contemporary culmination of metaphysical thought and practice in the regime of technology and totalizing "enframing [*Gestell*]."

One decisive basis for the kind of realism I defend here is the chain of consequences following from Cantor's discovery of the transfinite, and from its problematic accessibility to mathematical thought, up to and including Gödel's incompleteness results. In particular, I shall argue, these consequences offer to challenge the traditional conception of the human as an essentially *finite* agent of *representational* thought by locating the implications of a point at which thought touches on a "real" that is impossible, without contradiction, for such an agent. Through formal reflection, this "real" is indicated, not only as a positive structure of excess, but also as a point of impasse: the point at which a finitely constructed formalism reflexively captures its own specific structural limits, as well as those of the finitely constituted capacities it embodies. Basing itself on this, the metaformal realism I shall develop more fully here might be formulated, referring to the Lacanian motto according to which "the Real can be inscribed only on the basis of an impasse of formalization," as a realism *of* the "Real" in the sense of this motto.[4] The "Real," in this sense, is thought as one of the three "registers" of psycholinguistic development; in Lacan's system, it is both an inherent limit-point of structure and an obscurely constitutive underside for both of the other two "registers," the Imaginary and the Symbolic. As such, it structurally points to a necessarily displaced or "barred" position in relation to what Lacan characterizes as the "thing" or the "object small a."[5] Thought in this way, the "Real" is to be sharply distinguished both from "reality" in the sense of actuality and from any realm, regime, or domain of actually existing objects. But as Lacan himself occasionally suggests, the problematic of "access" to the Real, at the structurally necessary point of impasse which, for him, formally constitutes the subject's place in the order of the Symbolic, is essentially related to the "ontological" problematic of the "place" of being as such in relation to the factical life and language of the beings we are. As I shall argue here, this problematic, first developed (in *Being and Time*) as that of the constitutive structure of the kind of entity—Dasein—that is ontic-ontological, and later (after the mid-1930s) as that of the truth of being that comes to light as time, both suggests and requires the metaformal realism that I shall defend on partially independent grounds.

To arrive at the disjunctive conclusion that he draws in the lecture, Gödel draws on a concept central to twentieth-century inquiry into the foundations of mathematics, that of a "finite procedure." Such a procedure is one that can be carried out by a system governed by well-defined and finitely statable rules, a so-called "formal system." As Gödel points out, there are several rigorous ways to define such a system, but they have all been shown to be equivalent to the definition given by Turing of a certain specifiable type of machine (what has come to be called a "Turing machine").[6] The significance of the investigation of formal systems for research into the structure of mathematical cognition and reality lies in the possibility it presents of rigorously posing general questions about the capacities of such systems to solve mathematical problems or prove mathematical truths. For instance, one can pose as rigorous questions those of (i) whether such a system is capable of proving *all* arithmetic truths about whole numbers; and (ii) whether such a system is capable of proving a statement of its own consistency. Notoriously, Gödel's first and second incompleteness theorems, respectively, answer these two questions, for *any* consistent formal system capable of formulating the truths of arithmetic, in the negative: given any such system, it is possible to formulate an arithmetic sentence that can (intuitively) be seen to be true but cannot be proven by the system, and it is impossible for the system to prove a statement of its own consistency (unless it is in fact *inconsistent*).[7]

Gödel's argument from these results to his "disjunctive conclusion" in the lecture is relatively straightforward. The first incompleteness theorem shows that, for any formal system of the specified sort, it is possible to generate a particular sentence which "asserts" that it, itself, cannot be proven within the system. Thus, assuming the system's consistency and assuming that each sentence is either true or false, we can assert that the sentence in question is undecidable by means of the system, in the sense that neither it nor its negation can be proven. As it "asserts" its own unprovability, which we can *thereby* see (from a metasystematic perspective) to be a truth, it thus appears that there is a truth which cannot be proven within the system.[8] Mathematics is thus, from the perspective of any specific formal system, "inexhaustible" in the sense that no such formal system will ever capture *all* the actual mathematical truths.

Of course, given any such system and its unprovable Gödel sentence, it is possible to specify a *new* system in which *that* truth is provable; but then the new system will have its *own* unprovable Gödel sentence, and so on. The question now arises whether or not there is some formal system which can prove *all* the statements that *we* can successively see to be true in this intuitional way. If *not,* then human mathematical cognition, in perceiving the truth of the successive Gödel sentences, essentially *exceeds* the capacities of all formal systems, and mechanism (the claim that human mathematical cognition is, or is capturable by, a formal system) is false; this is the first alternative of Gödel's disjunction. If *so,* however, then there *is* some formal system that captures the capacities of human mathematical thought.[9] It remains, however, that there will be statements that are undecidable for this system, including the statement of *its* consistency, which is itself simply a (negatively existentially quantified) arithmetical statement. Thus it is impossible, on this alternative, simultaneously to identify a set of underlying principles on which actual mathematical cognition is based and to claim that these principles are both consistent and capable of deciding all mathematical problems. In this case there are thus classes of problems that cannot be solved by any formal method we can show to be consistent *or* by any application of our powers of mathematical cognition themselves. These are, then, apparently well-defined (classes of) problems which will remain unsolvable by any definite method, now and for all time.

We can further specify the underlying issue, and move closer to its philosophical significance, by noting that, by Gödel's second theorem, the undecidable Gödel sentence for each system is equivalent (even within the system) to a statement, within that system, of its own consistency. As Gödel emphasizes, it is (given classical assumptions) an implication of the *correctness* of any system of axioms that we might adopt for the purposes of arithmetic demonstration that the system be consistent. But then it is an implication of the second incompleteness theorem that if we are in fact *using* a specific (and consistent) formal system to derive all the mathematical truths (that we know) we cannot know that we are. For if we could know this—that is, if we could know the truth of the assertion of the consistency of the system—we would thereby know a mathematical truth that cannot be derived from that system.

Accordingly, as Gödel says, it is

> impossible that someone should set up a certain well defined system of axioms and rules and consistently make the following assertion about it: All of the axioms and rules I perceive (with mathematical certitude) to be correct and moreover I believe they contain all of mathematics.[10]

Thus if a system is (knowably) consistent it is, by that token, *demonstrably* incomplete; if it is complete, it is not consistent (and hence we cannot know it to be correct). Accordingly, on the assumption that we are in fact using a finite procedure to demonstrate mathematical truths, the assumption of the consistency of the system we are actually using is shown to be *essentially* unsecurable in any way that is itself consistent with our (in fact) using (only) that system at all.

Again, by considering the question of the axiomatization of mathematics, we can see how the issue is connected to the problem of the accessibility of the infinite, and the higher levels of infinity. Specifically, in order to axiomatize arithmetic set-theoretically without contradiction, it is necessary to introduce axioms in a step-by-step manner, and in fact, as Gödel suggests, this process can be continued infinitely: thus,

> instead of ending up with a finite number of axioms, as in geometry, one is faced with an infinite series of axioms, which can be extended further and further, without any end being visible and, apparently, without any possibility of comprising all these axioms in a finite rule producing them.[11]

The successive introduction of the various levels of axioms corresponds to the axiomatization of sets of various order types; in each case the introduction of a new level of axioms corresponds to the assumption of the existence of a set formed as the *limit* of the iteration of a well-defined operation. But each axiom "entails the solution of certain

Diophantine problems, which had been undecidable on the basis of the preceding axioms." In particular, according to a result that Gödel had achieved in the 1930s, the consistency statement for any given system of axioms can be shown to be equivalent to a statement asserting the existence of integral solutions for the variables of a particular polynomial equation.

Since consistency is undecidable within the system itself, so is the problem of the truth-value of the statement concerned, but it becomes decided in a stronger system which adds, as a new axiom, a statement of the former system's consistency (or something equivalent to this).[12] But since the problem of the truth of the statement about the solutions to a polynomial is itself simply a number-theoretical problem, it follows that each particular system, if it *is* consistent, cannot solve some mathematical problem; and that *if* human cognition is equivalent to some *particular* system then there is some problem of this form (equivalent to the statement of its own consistency) that *it* cannot solve either. This is then an "absolutely undecidable" problem. If, however, there is no formal system to which human cognition is equivalent, then for *any* specified machine the mind can prove a statement which that machine cannot, and accordingly "the human mind . . . infinitely surpasses the powers of any finite machine."[13]

2. Thinking and Being: The Four Orientations of Thought

I term an *orientation of thought* a schema of the relationship between thought and being, identified by the form in which it modulates this relationship according to the "metaformal" ideas of consistency, completeness (or totality), and the reflexivity involved in thinking the totality from a position within it.[14] Since each orientation determines that thought is thus related in some formally specifiable way to the infinity of thinkable beings, each one can be specified in terms of the specific figure of the infinite that it propounds. This allows us to specify the four orientations of thought according to these figures of the infinite, as well as the way they are produced by the specific relationship of the metaformal ideas in each case (figure).

First, the *onto-theological orientation*, marked in traditional thought since Aristotle, supposes the infinite as a transcendent Absolute, complete and consistent in itself but inaccessible as such to finite, human cognition. (I use "onto-theology," with one hyphen rather than two, to designate this specific orientation, by contrast with "onto-theo-logy" [two hyphens], which is synonymous with what Heidegger usually calls "metaphysics" and which comprises both the constructivist orientation *and* onto-theology [in the narrower sense].)

Second, the *constructivist* orientation, original with Kant, propounds the infinite as always only *potential* in order to delimit the totality of what can be *known* according to regulative protocols or coherent criteria of finitude, while the totality of *what is* (in the Kantian context, the total world of things in themselves) remains jointly *consistent* and *complete* (though as such inaccessible to finitely constituted knowledge).

But beyond these initial two orientations, there are the two opened by the Cantorian discovery of the transfinite as irreducibly plural and as permitting the inscription of the whole or totality only on pain of unavoidable paradox. These are the "generic" and the "paradoxico-critical" orientations, and they directly correspond, as well, to the two options left open by Gödel's own disjunctive conclusion in the 1951 lecture. On the first of Gödel's options (the one on which the human mind can recognize the truth of an infinite number of propositions beyond the powers of formal systems to demonstrate), this power witnesses an essential *incompleteness* of any finitely determined cognition. This is seen as motivating or requiring a correlative capacity on the part of human thought, rigorously following out the mandate of consistency, to traverse the infinite consequences of truths essentially beyond the reach of any such finite determination. On the second of the options (that of problems forever unsolvable by formal *or* "human" means), the essential *undecidability* of any such system shows, rather, that the *consistency* of any finitely determined procedural means to truth is *undemonstrable* by those or, indeed, *any* means.

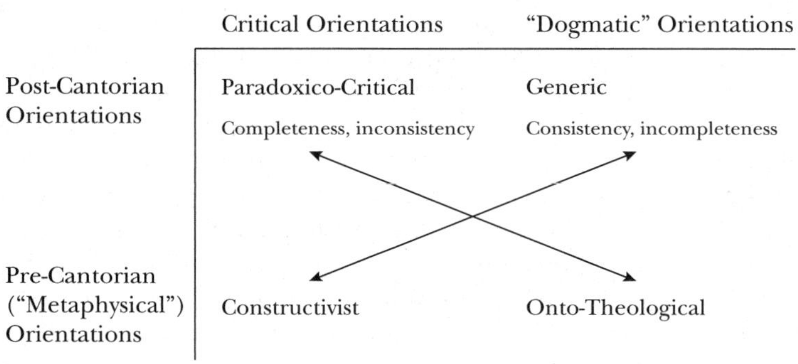

The two "post-Cantorian" orientations can be distinguished from the two "pre-Cantorian" on the basis of the specific relationship each envisions between thought and being, in connection with the idea of the *totality* of *what is* (in Heidegger's jargon, *das Seiende im Ganzen*) as such.

As I use the phrase here (and, I believe, as Heidegger uses it as well), this "totality" does not mean "all that (contingently) exists." At any given time, of course, it is possible to use language to designate many entities that exist in this sense, and also others that do not or no longer do. But the phrase *das Seiende im Ganzen* takes in both of these categories and, more broadly, anything that "is" in any way at all. This includes anything that can be characterized in any way whatsoever: the (nonexistent) centaur, for instance, "is" in this sense, for it has specific and determinate properties (having the head of a man and the torso of a horse) in terms of which it may be characterized.[15] The question to be raised here—and which is treated differentially by means of the four orientations—is about the being and properties of the totality of everything that is, in this sense. In particular, the totality will be *complete* if, and only if, *there is nothing it does not include*; it will be *consistent* if all truths about it are themselves consistent.[16]

Both the onto-theological and constructivist orientations can be distinguished from the "post-Cantorian" ones in that they presuppose, though on different grounds, that the totality exists and is both *complete* and *consistent*. In particular, whereas the onto-theological orientation assumes an infinite or transcendent consistent totality that is in itself never completely accessible to "finite" human cognition, the constructivist orientation constitutively involves the assumption of a thinkable and consistent totality that is then *subsequently* regulated, limited, or constructed by subjective procedures, activities, or forms. Because both of the "pre-Cantorian" orientations develop their respective conceptions of the thinkability of being in terms of the thought of a consistent and complete totality, they both amount to primarily *ontic* orientations toward what they figure as the totality of present-at-hand (*Vorhanden*) beings. On the one hand, onto-theology, understanding this totality as a complete and consistent whole quite independently of the capacities or activities of subjects or thinkers, captures many of the forms of realism that have been articulated in the metaphysical tradition. On the other, constructivism, in lodging the combination of consistency and completeness in the field of application of the constitutive activities or abilities of a subject of experience, dialectical self-recognition, or linguistic institution, encompasses most forms of subjective, transcendental, speculative, and linguistic idealism or antirealism.[17]

If, however, the set-theoretical and semantic paradoxes already indicate the untenability of the conjunction of consistency and completeness that both "pre-Cantorian" or "metaphysical" orientations assume, then the orientation of thought toward being can subsequently only be conceived in one of the two "post-Cantorian" ways: either in terms of the generic orientation, which preserves consistency while sacrificing completeness (thus maintaining the consistency of rules of inference while sacrificing the existence of a total world to which they apply), or the paradoxico-critical orientation, which develops the thought of the *constitutively inconsistent* whole. As we shall see, the impossibility of the joint ("pre-Cantorian") assumption of consistency and completeness receives further confirmation, and is put on a farther-reaching basis, in relation to the constitutive ideas of practice, method, and procedure, by means of Gödel's and Turing's undecidability results. These results jointly bear witness to an essential undecidability at the limit of all possible procedures, methods, or activities of human (or any) rationality, insofar as these procedures, methods, and activities are determined by rules in any sense. The formally indicated limit of the powers of the thinking subject then becomes visible as the closure of the metaphysical assumption of complete and consistent totality, and thereby of the whole historical epoch that Heidegger understands as that of the "metaphysics of presence."

Because both of the "pre-Cantorian" orientations deploy their understanding of the relationship between thinking and being on the basis of the presupposition of the joint completeness and consistency of the *ontic* totality, they are equally (and in parallel fashion) overcome by the thought of the ontological difference. As Heidegger recurrently emphasizes in his discussions of the metaphysical tradition, the various configurations and approaches of metaphysics, beginning with Parmenides, always have in view the ontic *totality* of present-at-hand beings and always think this totality on the basis of the assumption of its joint *consistency* and *completeness*. This assumption of consistency and completeness remains just as much in force in what I have called the constructivist orientation, where the complete and consistent totality of beings is thought, on the one hand, as the totality of *objects* conditioned by the constructive or limitative activities of a subject and on the other as the thinkable but unknowable totality of *things in themselves*. The same orientation is maintained, albeit sometimes in a less overtly subjectivist fashion, by (most forms of) intuitionism, verificationism, and antirealism in the philosophy of mathematics. What is crucial to the constructivist orientation as such is not simply the idea of subjective limits or constraints, but rather the attempt to understand what can be "given" for content-bearing thought as delimited from outside by means of some specific limitative criterion or another (what is knowable or what is ultimately intuitable, what is temporal, what can be given in a finite intuition, etc.).[18] What Heidegger calls the history of "onto-theo-logy" (in the broader sense in which it includes as well the constructivist orientation) is identical with the history of this assumption of conjoint completeness and consistency in the totality of beings, and so the ontological displacement that it undergoes through Heidegger's thinking of the ontological difference and its radicalized consequences also provides ontological terms for the transition from the "pre-Cantorian" to "post-Cantorian" orientations.

The idea of an essential disjunction and forced choice, in the wake of metamathematical results, between recent and contemporary positions that maintain consistency but sacrifice completeness and those that maintain completeness along with inconsistency or paradox, was first suggested to me by John Bova in conversations I had with him in 2008 and 2009.[19] It was in considering the implications of this disjunction that I developed the framework of the four orientations of thought (as explicated in *The Politics of Logic*, chapters 1 and 9). There I did not consider in any detail the relationship between Bova's "metalogical" difference or duality of completeness and consistency, on the one hand, and the ontological difference, on the other. However, in light of the present considerations, it appears that the metalogical results can indeed be seen as verifying, when situated in terms of the four orientations, a structural convergence of the two kinds of difference. In particular, the metalogical difference that produces the two post-Cantorian orientations and their distinction from one another—that between consistency with incompleteness and totality with inconsistency—can *also* be seen as a figure or articulation of the ontological difference, once the "relationship" between being and beings is *itself* understood in terms of the metalogical results.

How should we then see the implications of an attitude of overall ontological realism with respect to each of the four orientations? In the most general terms, the argument is as follows. The maintenance of realism with respect to the ontological difference—the maintenance, that is, of a realist attitude with respect to statements involving the difference between being and beings—yields the necessity of metalogical difference: that is, of the *disjunction* of the positions of completeness with inconsistency (paradoxico-critical) *or* consistency with incompleteness (generic).[20] It thereby also renders inaccessible either of the two pre-Cantorian orientations, which both presuppose (in different ways) the combination of completeness with consistency. Given this, it remains possible, however, to maintain *either* of the two post-Cantorian orientations, each of which witnesses the deadlock between completeness and consistency, but each of which resolves it in a different way. There are then further considerations, relative to particular questions and problems, which might be entered in favor of one of the post-Cantorian orientations over the other.[21] But in the present context it is less important to argue for one over the other, or indeed to determine which of the two post-Cantorian orientations Heidegger "himself" occupies, than to see how the ontological difference can itself be articulated in both ways and how the two articulations complement one another within a broader posing of the ontological question and within the development of the "history of being" it determines.

First, there is the generic position, which maintains consistency by situating thought with relation to an always-external matter that insists in it at the point of the truth it properly reveals. Second, there is the paradoxico-critical one, which sees the closure of the totality of beings only in the light of a fundamental structure of paradox that always again arises at the boundaries of any complete determination of it. The first points to that which, in determinate total configurations of beings as a whole, points beyond their determining (ontic) figure to their ultimate dispensation in "being itself," while the second points to the structurally necessary paradox which, in every such configuration, undermines the consistency of its basis and organizing structure. Together, they articulate jointly and formally the more general structure underlying the epochal determination of the successive figures of the being of beings. For the same reason, they articulate as well the formal structure of the contemporary historical "closure" of onto-theo-logy, the situation in which its possibilities of defining beings are definitively exhausted.

It is true that Heidegger does not generally distinguish, in his own accounts of the history of metaphysics up to the present, the two "pre-Cantorian" orientations, which characteristically neglect both ontological and metalogical difference. This unclarity in Heidegger's own retrospective discussion is one root of the contemporary interpretive tendency which, while grasping to some extent the terms and implications of Heidegger's critique of constructivism, nevertheless continues to attribute to him some form of the thesis that limits the "accessibility" of being as such to the powers of a thinking (individual or collective) subject. Heidegger then can only appear as a kind of post-Kantian antirealist, and of course it is not difficult to locate in his writing, particularly in *Being and Time*, the conception of a structural "transcendence" which, as in Kant, is then thought to underlie the constructive relationship of something like a "human" subject—albeit now the living and factical subject of "embodiment" and practices rather than the "intellectual" or "worldless" subject of Descartes, Kant, and (it is supposed) Husserl—to entities or their being. The conception has the additional feature that it conforms with a prevalent (and essentially constructivist) conceit of contemporary belief, according to which, if there is no "ultimate" theological referent to hold together the totality of the world as an intellectually thinkable unity, such access to being as is possible "for us" must instead be irreducibly mediated by the plural variety of bodies, languages, and situated cultures.[22]

As we shall see in this chapter and the next, however, given an explicit identification of the metalogical issues at stake in the four orientations, metaformal realism points to a very different conception of the ontological problematic that is already suggested by Heidegger in *Being and Time,* albeit only developed fully in the course of his radical critical encounter with Kant in *Kant and the Problem of Metaphysics.* On this conception of the ontological problematic, it deconstructs the idea of a constitutively essential basis for ontology in "human" thought, language, culture, embodiment and practices just as thoroughly, and on substantially the same basis, as it does the "theological" appeal to the transcendent Absolute. As such, this conception, as we shall see, substantially underwrites the possibility of a *realist* picture of the ontological difference and of the ontological structure of time.

Put in terms of the questions which Gödel's work on undecidability and incompleteness addresses, the issues of completeness, consistency, and totality are naturally understood as issues about the structure and powers of theories (that is, axiomatic formal systems) of a certain sort.[23] In particular, and in these terms, as we have seen, Gödel's two "incompleteness" theorems demonstrate that theories of this sort, (i) are either inconsistent or, if consistent, incomplete (in the sense that there is a sentence such that neither it nor its negation is derivable, and so—assuming at least bivalence—at least one nonderivable sentence that is *true*); and (ii) cannot prove a statement of their own consistency. But if we may take these results, as Gödel himself does, to have a more general significance for the question of the relationship of thought to the objects or truths it bears upon, we may naturally put them in terms of their implications for realism or antirealism about these objects or truths (including the *totality* of objects or truths of the kind treated by a particular theory).[24]

Much of the discussion in the extant philosophical literature over the broader implications of Gödel's theorems has been directed toward the question of the truth or falsity of *mechanism,* or of whether the mathematical thought of an individual subject, or perhaps of the whole community of mathematicians, can "in fact" be captured by some formal system. It is not clear, despite this discussion, that the problem can be well posed, and even if it can, it appears possible on the basis of Gödel's results only to draw a "disjunctive" conclusion of the same form as the one he draws explicitly in the 1951 lecture: *either* human minds cannot be captured by formal systems and so have (somewhat mysterious) powers to perceive or demonstrate truths beyond the scope of finite, algorithmic procedures *or* "minds are machines" after all and there are problems that are *absolutely* unsolvable by either.[25] But beyond the issue of mechanism, it is also possible to see the upshot of Gödel's "disjunctive conclusion" in the lecture as bearing relevance to broader and different philosophical issues.[26] In particular, it points to a distinctive and nonstandard, but comprehensive position of *realism,* what I shall call *metaformal realism.*[27] For this realism, the decisive issue is not, primarily, that of the reality of "mathematical objects" or the possibility of understanding them as determinate independently of the routes of access to them (epistemic or otherwise) involved in the exercise of our human capacities. It is, rather, that both terms of Gödel's disjunction capture, in different ways, the structural point of contact *between* these capacities and what must, on *either* horn of the distinction, be understood as an infinite *thinkable* structure determined quite independently of anything that is, in itself, finite. Thus, each term of Gödel's disjunction reflects the necessity, given Gödel's theorems, that any specification of our relevant capacities involves their relation to a structural infinity about which we must be realist: that is, which it is not possible to see as a mere *production* or *creation* by means of these capacities.

On the first alternative, this is fairly obvious. If human mathematical thought can know the truth of statements about numbers which are beyond the capacity of *any* formal system to prove, then the epistemic objects of this knowledge are "realities" (i.e., truths) that also exceed any finitely determinable capacity of knowledge.[28] It does not appear possible to take these truths as "creations" of the mind unless the mind is not only credited with *infinite* creative *capacities,* but understood as having actually already *created* all of a vastly infinite and in principle unlimitable domain. But on the second alternative, it is equally so. If there are well-specified classes of mathematical problems that are not solvable, as classes, by any means whatsoever, neither by any specifiable formal system nor by human cognition itself, then *these problems* must be thought of as realities determined quite independently of our capacities to know them (or, indeed, to solve them).[29]

The actual underlying reason for the realism that appears forced upon us on either alternative is the phenomenon Gödel describes as that of the *inexhaustibility* of mathematics. This inexhaustibility can be seen as resulting from the possibility of considering, given any well-defined ordinal process, its infinite limit (or totality). On the first alternative, this inexhaustibility yields a structurally necessary *incompleteness* whereby each finite system by itself points toward a truth that it cannot prove but which is nonetheless, by this very token, accessible to thought. On the second, it yields an equally necessary *undecidability* which leaves well-specified mathematical problems unsolvable by any means whatsoever (finitely specified or not).[30] The forms of realism involved in the two positions are somewhat different. The orientation underlying the first disjunct corresponds, as I argued in *The Politics of Logic,* to a realism of *truth beyond linguistic sense,* a position that affirms the infinite existence of truths and the infinite genericity of our dynamic insight into them beyond any finitely specifiable language or its powers. By contrast, the realism of the second is a realism of *sense beyond linguistic truth,* affirming the existence of linguistically well-defined *problems* whose truth-value remains undecidable under the force of any powers of insight whatsoever. But in either case, reflective thought about our capacities must reckon with the consequences of their structurally necessary contact with an infinite and inexhaustible reality essentially lying beyond the finitist determination of the capacities of the human subject or the finitely specifiable powers of its thought. In this way, the consequences of Gödel's theorem, however we interpret them, engender a structurally necessary realism about the objects of these powers that is the strict consequence of the entry of the infinite into mathematical thought.[31]

3. Dummett: Realism and the Basis of Sense

In contemporary philosophical discourse, no project has done more to illuminate the issue of realism and its underlying formal determinants than Michael Dummett's. In a series of articles and books beginning in 1963 with the article "Realism," Dummett has suggested that the dispute between realism and antirealism with respect to a particular class of statements may be put as a dispute about whether or not to accept the principle of *bivalence* (i.e., the principle that each statement is determinately true *or* false) for statements in the class concerned.[32] Though this issue yields differing consequences in each domain considered, the acceptance of bivalence generally means the acceptance of the view that all statements in the relevant class have truth-values determined in a way in principle independent of the means and methods used to verify them (or to recognize that their truth-conditions actually obtain when they, in fact, do so). The antirealist, by contrast, generally rejects this view with respect to the relevant class. Dummett did not envisage that this comprehensive framework would or should support a single, *global* position of metaphysical "realism" or "antirealism" with respect to all domains or the totality of the world. Rather, his aim was to illuminate the different kinds of issues emerging from the traditional disputes of "realism" and "idealism" in differing domains by submitting them to a common, formal framework.[33] From the current perspective, however, it is just this aspect of formal illumination that is the most salutary feature of Dummett's approach. For by formally determining the issue of realism with respect to a given domain as one turning on the acceptance or nonacceptance of the (meta)formal principle of bivalence with respect to *statements*, Dummett points toward a way of conceiving the issue that is, in principle, quite independent of any *ontological* conception of the "reality" or "ideality" of *objects* of the relevant sort.[34] In particular, it is in this way that Dummett avoids the necessity to construe realism and antirealism in *any* domain as involving simply differing attitudes toward the ontological status of its objects (for instance, that they are "mind-independent" or that, by contrast, they are "constituted by the mind"). What this witnesses, *along with* what I have called metaformal realism, is the possibility of a purely formal and reflective determination of realism that connects its stakes directly to those of the truth of claims, thereby instantly short-circuiting the laborious and endlessly renewable dialectic of the "subject-object" relationship, or of that of "mind" to "world."

It is important at this point also to note carefully that bivalence, in the sense thus defined, need not exclude the possibility of *contradiction*. As the principle that each statement in a given domain is *either* true *or* false, it does not and need not exclude the possibility that a statement may be *both*. Equivalently, though bivalence requires the semantic analogue of the law of the *excluded middle* (A v ~A) for any statement in the relevant domain, it thus does not necessarily require the analogue of the law of *noncontradiction* (~[A&~A]). If the formal requirement of bivalence in Dummett's sense can indeed be identified as formulating the requirement of realism, it is then significant that putting things in this way allows the possibility of realism *about* contradictions: that is, about the situation in which, for a given A, both A and ~A are true.[35] This proves especially important in considering the formal options available for reflecting the consequences of metaformal results, especially those about the limits of totalities, the structure of reflexivity, and the consequences of ontic-ontological difference.

In the 1973 article "The Philosophical Basis of Intuitionistic Logic," Dummett considers the question of what rationale might reasonably serve as a basis for replacing classical logic with intuitionistic logic in mathematical reasoning (hence, in his framework, for replacing realism with antirealism with respect to mathematics).[36] In fact, Dummett holds, there are just two lines of argument that could plausibly be used to support the replacement. The first turns on the idea that "the meaning of a mathematical statement determines and is exhaustively determined by its *use*." Beginning from this assumption, it is plausible to hold that any difference between two individuals in their understanding of mathematical symbolism would have to be manifest in observable differences of behavior or capacities. The second turns on the thought that what it is to *learn* mathematical reasoning is to learn how to *use* mathematical statements (i.e., when they are established, how to carry out procedures with respect to them, how to apply them in nonmathematical contexts, etc.).[37] On either assumption, Dummett holds, *since* meaning is exhausted by use (in one way or the other), we cannot claim that a notion of truth, understood classically as imposing bivalence on all mathematical statements independently of the use we actually make of them, can any longer serve as the "central notion" for a characterization of the meanings of mathematical statements. In place of the classical notion of truth, he suggests, we must thus substitute a notion grounded in the practices of which we have actually gained a mastery, and in particular one which captures that "a grasp of the meaning of a statement consists in a capacity to recognize a proof of it when one is presented to us."[38]

Dummett thus presents the best route to the adoption of intuitionistic logic in mathematics as motivated by considerations very different from those that motivated arguments to the same conclusions for classical intuitionist thinkers such as Brouwer and Heyting. In particular, as Dummett points out, whereas intuitionism was motivated for those thinkers primarily by the requirement that mathematical objects be present or given in subjective, private experience, Dummett's arguments turn on what is in some ways the exactly opposite idea, namely that of the mastery of a socially learned and publicly evident *intersubjective practice*. In fact, Dummett suggests, against the views of the early intuitionists, that there is *no* plausible route from the idealist view that mathematical entities such as natural numbers are "creations of human thought" to the application of intuitionistic logic, unless we are prepared to adopt a very severely restricted (and implausible) view of mathematical practice (including rejecting unbounded quantification over all numbers, etc.).[39] For this reason, Dummett suggests as well that there is no good reason to think that any successful argument for antirealism in mathematics can turn on considerations bearing simply on the supposed ontological peculiarities of the "mathematical domain." Both of the reasonable arguments that one might make turn, instead, on considerations about the link between meaning and linguistic use. As such, they have nothing special to do with mathematics and appear to be applicable much more broadly, to any number of classes of sentences "about" widely differing kinds of things.

Because the question of realism versus antirealism in any domain thus turns, for Dummett, on the question he puts as that of the *provision* of sense (rather than, for instance, the ontological status of objects, or their dependence on creative acts of the mind), it is to be answered by way of an answer to the question of how the *meanings of statements* about entities in the domain are generally learned or manifested. But in the current context, this is none other than, again, the question of the way that the *infinite* becomes available on the basis of a *finite* procedure. For the intuitionist (and, by analogy, the antirealist more generally), it is possible to establish the existence of an object only if it can be shown to result from its actual construction in a finite number of steps or from a finite, constructivist proof (i.e., one that does not involve reasoning over arbitrarily complex infinite totalities). By contrast, for the formalist (realist), all that is needed is to show that it is possible to refer to the object within a specified formal system. And it is just here, with regard to the specific question of what is involved in the learning and carrying out of a finite procedure as such, that the possibility of metaformal reflection of the sort that I have portrayed Gödel as engaging in proves decisive. For Gödel's own incompleteness theorems, as we have seen, directly embody a rigorous metaformal consideration of the range and capacities of formal systems (in Hilbert's sense and related ones). Specifically, Gödel's first theorem shows that for any such system, there will be a number-theoretical sentence that is beyond its capacity to prove or refute, and the second theorem shows that no such system can prove its own consistency (assuming that it *is* consistent). In this way, Gödel's results render Hilbert's formalist conception of finite procedures unsuitable for anyone who wishes to base upon it the realist position that the statements of number theory have determinate truth-values, quite independently of our ways of verifying them. But on the other hand, in invoking under the heading of the "inexhaustibility" of mathematics an essential reference to a reality that marks the point of impasse of any given finite procedure, Gödel's argument shows the intuitionist strictures against reference to truth beyond the possibility of intrasystematic proof to be untenable as well.

What, then, of Dummett's own arguments for antirealism with respect to mathematics, as well as in various other domains? The general form of this argument is the one we have already seen. On the assumption of the identity of meaning and use that Dummett attributes to the late Wittgenstein (correctly or incorrectly), sense must be provided or established for any range of sentences by means of the establishment and learning of a (public, intersubjective) *practice*. On Dummett's various arguments, this makes it incoherent, in a variety of domains, to suppose that sense could have been provided or determined independently of procedures for verifying or recognizing truths in those domains. According to Dummett, it is for this reason, for instance, that one must be antirealist about descriptions of "private" experience, and these considerations at least suggest antirealism about the past (though, as Dummett admits, matters are more complex here owing to the internal complexity of the notion of the [current] verification of [past] events).[40] The general argument goes through, as we have seen, on the assumptions that (i) sense is provided by means of a *practice* whose institution essentially *involves* laying down various well-defined procedures of verification and (ii) this provision of sense is *not intelligible except by means* of the specification and establishment of these procedures.

However, both of these assumptions are at least contestable in a context where (as I have argued with respect to the kind of metaformal reasoning Gödel applies) sense appears to be "provided" through finite instances of teaching and learning, but *in such a way* as to essentially *outstrip* any description of the outcome of these instances as the internalization of finite *procedures.* And truth here is demonstrated, not by any simple application of established verification procedures, but on a constitutive *reflection* on their scope and limits that is itself *irreducible* to any antecedently given procedure. More generally, since Dummett's arguments turn on the thought that sense, if it is to be determinate, must first be "provided" by the human activity of instituting or learning such procedures, it can be resisted where we have good independent reason to consider sense to be "given" in a way that essentially outstrips such procedures. Such an independent reason is provided, as I have argued, by the metaformal reflection that underlies Gödel's results and thereby demonstrates what he calls the "inexhaustibility" of mathematics, and more generally by the problematic accessibility of the infinite and transfinite to thought that is broadly witnessed in the results of Cantor, Gödel, and Turing.

More narrowly, Dummett himself has, at least sometimes, suggested an argument from Gödelian incompleteness to antirealism about mathematics. For instance, in the 1963 article "The Philosophical Significance of Gödel's Theorem," Dummett considers a common model-theoretic explanation of the first incompleteness theorem. According to this common explanation, since the Gödel sentence, U, for a particular system can be recognized (in some sense) as true, but cannot be proven in that system, the system must have some *models* in which U is true and others in which it is false. Thus, to say that U is *true* is not to say that it is true in all models, but only that it is true in (as we may be inclined to put it) the "intended" model—that which includes all and only the "actual" natural numbers. As Dummett notes, for this explanation to succeed, it must then be the case that we can have a "quite definite idea" of the mathematical structure of the natural numbers, but one which is not fully specified by any formal system. On the view, it is to this "intuitive" idea that we must refer when recognizing the actual "truth" of the statement U. However, it must also be impossible to characterize it by means of a formal system, or indeed by *any* finitely statable stipulation of the *assertions* we are prepared to make about natural numbers.[41] This is because any such finite stipulation, which is intended to capture completely and exclusively what is here viewed as our "intuitive" conception of the natural numbers, will itself produce a new Gödel sentence which we can see to be true, but which cannot be proved by the system, and thus its claim to have fully captured our intuitive conception will fail.

As Dummett recognizes, this line of thought threatens the generality of the thesis that meaning is to be explicated in terms of use. For it implies that, as Dummett puts it, "we have a certain, quite definite, concept" of the natural numbers, but one which nevertheless "cannot be fully characterized just by the fact that we make certain assertions about it."[42] As Dummett argues, there are good reasons for suspicion about the tenability of any such conception of an intuitive grasp that is completely definite and yet cannot be manifest in any finite description of use.[43] More generally, although Dummett acknowledges that it is not immediately clear how to construe the thesis that all explanation of meaning must be reducible to the finite description of patterns of use, to give up on this thesis is to "to fall back on a conception of meaning as something unanalysable; the notion of meaning will then be made to play a crucial role in explanations, but becomes itself quite incapable of explanation."[44] The proponent of the idea of an internal, intuitive grasp of the concept of the natural numbers, which cannot be manifest in any finite description of use, is in effect committed to such a "retrograde" and mystifying view about meaning, at least in this case.

Dummett grants that, on any account, if we consider the Gödel sentence for a particular system to be a number-theoretic truth, it will be necessary to acknowledge that for any such system "there is . . . something which leads us to recognize the [Gödel sentence] as true, and which therefore goes beyond the characterization of the natural numbers which is embodied in the formal system."[45] In particular, and significantly, it is only on the assumption of the *consistency* of the system in question that we can conclude that the relevant sentence U is in fact true. This assumption, Dummett notes, goes significantly beyond what must be assumed simply to accept any formal proof in the system as establishing its conclusion in general (which requires only the truth of the premises and the correctness of the reasoning employed in each of its finitely many steps). In particular what is assumed here must be strong enough to establish the truth of *all* of the statements A(0), A(1), A(2), . . . , for the suitable predicate A(x). And although there are familiar consistency proofs—for instance, proofs by induction on the length of formulas of the type first developed by Gentzen—for *specific* formal systems, what is really at issue here, as Dummett points out, is the basis for our *general* conclusion, from

Gödel's result, that *any* axiomatizable formal system must fail to capture *some* number-theoretic truths. But any formal demonstration of the consistency assumption underlying this general result that followed the familiar pattern for consistency proofs would have to embody induction, not just over the specific properties defined *within* a particular formal system, but over the totality of *all* "well-defined" properties of natural numbers.

But according to Dummett, it is just this totality that we cannot have access to in any general and formally definable way. For *given* any formulation of a system, S, capable of expressing certain properties, we can *always* define new properties—for instance, the property of being true-in-S itself—that are not expressible in S itself (again, on the assumption of the consistency of S). It is in this, Dummett suggests, that we should see the real philosophical significance of Gödel's theorem as consisting. In particular, it shows that the concept of a ground for asserting that something is true of all natural numbers is itself "indefinitely extensible": any definite characterization of this concept (such as is formulated in a particular formal system) will permit a "natural extension of this characterization, which yields a more inclusive concept."[46] There is therefore no possibility of capturing in a finite description the *totality* of our criteria for the truth of number-theoretic statements. He further suggests that we may conclude from this that our concept of "natural number," since it indeed involves the idea of the validity of induction with respect to *all* well-defined properties, is one that is "inherently vague," in that criteria for the assertability of truths about *all* natural numbers cannot be uniquely specified because these criteria are themselves indefinitely extensible. In this way, antirealism can be maintained—or at least the door to it is left open—by means of the thought that, being vague in this sense, these criteria are not determinate beyond the range within which they have *actually* (so far) been determined.

There are several points of contact between Dummett's argument and the metaformally realist interpretation of the significance of Gödel's results I have suggested here. First, of course, the phenomenon of *unlimited* extensibility that Dummett sees as integral to the correct interpretation of the results is substantially the same phenomenon that Gödel himself recognizes as the *inexhaustibility* of mathematics with respect to any finite formulation of it. Further, on the current position as well as on Dummett's, it is this essential inexhaustibility (understood also as demanding the interpretive metalogical choice between consistency and completeness) that captures the real implications of Gödel's result and characterizes the epistemological situation we are in with respect to the arithmetical properties of natural numbers. Moreover, on the present position as on Dummett's, it is essential to a proper characterization of this situation that it be described (in *some* sense) in terms of our *actual use* of our terms and concepts, rather than understood as committing us to some ultimately mysterious capacity for an intuitive total grasp of what could not *in any way* be expressed finitely in use. Finally, the metaformal realist agrees with Dummett in seeing the implications of this epistemological (and metalogical) situation, whatever they may be, as extending beyond the confines of arithmetic per se, and having a broader relevance in particular to the question of *how* sense is provided for assertoric sentences generally.

Nevertheless, from a realist perspective, a number of questions can and should be raised about Dummett's conclusion of an inherent "vagueness" in the concept of natural numbers. To begin with, in construing the phenomenon as implying irreducible vagueness in this sense, Dummett must take it for granted that there is *no* perspective available to us from which it is possible to have in view a legitimate and rigorous ground for drawing conclusions about the properties of natural numbers, if "properties" refers to *all possible* properties, rather than just those that are in fact expressed in some particular formal system. But the result of Gödel's reasoning, if taken generally, is arguably just such a conclusion: we can know from it that *no* (finitely expressible and consistent) formal system can establish all (of what we can recognize as) arithmetic truths. Put this way, the conclusion goes significantly beyond the mere observation, in the case of *each* formal system, that *that system* can be extended by introducing new properties not expressible in it. Rather (as Dummett in fact recognizes), if the point is to be put generally, it appears to *require* expression in terms of a conception of natural numbers that licenses the validity of induction with respect to *all* properties (not just those expressible in terms of any particular system). But if this is so, we must then have some kind of access to this *totality* of properties even in order to understand Gödel's result in its full generality at all.[47] Much the same point can be made, as well, with respect to Dummett's own conception of the basis for the indefinite extensibility he cites: to say that systems *can* always be extended in the regular or "natural" way suggested here is also, quite apparently, to imply that there is some (fully determinate) basis for their being extended in just this way rather than any other. It is not, for instance, simply "up to us" to decide *how* to extend them in each case. In doing so, in practice, we can in fact rely fairly confidently on the assumption that another reasoner will perform the extension in the same way I do, or at least that our results will not, themselves, be inconsistent with one another. This reliance witnesses our (indisputably real) confidence that we are not extending the system simply arbitrarily, but rather in view of what are *actually* the truths about natural numbers "themselves."

These considerations appear to motivate a realist, rather than antirealist, conception of the criteria by which the unlimited (though in fact *not* "indefinite") extension of particular formal systems can be carried out, as well as a realist construal of the topic of the reasoning that establishes this inexhaustibility *in general*. As Dummett argues, of course, what makes it difficult to maintain such a position of realism about the criteria for truth is that, by Gödel's own result (as well as Tarski's closely related ones), these criteria, as we actually use them, must apparently outstrip any *finite description* of them. But why should not both the real criteria for arithmetic truth, as well as the real basis for the generality of Gödel's result, be *formally indicated* in finite use, even if they are indeed never exhaustively *definable* by means of any finite *description* of the practical basis of use? On the metaformal realist position, as we have seen, the underlying structures at the unitary basis of truth and sense, although they necessarily elude complete definition in any one formal system or language (for instance, in any particular definition of "True-in-L" for any particular language L), are nevertheless capable of being *indicated* in their generality through the very metaformal considerations and results that also show the limitation of any such (finitely specified) definition itself. As thus indicated, these structures are also indisputably *real*: for instance, they are in no sense simply the product of human choice, decision, or convention. What is needed to verify the possibility of this realism—which is anyway (as I have argued in the last paragraph) essentially presupposed in the description of our concepts as extensible in the way Dummett suggests—is just an account of the way in which a reflection on usage—of the type that the metaformal results effectively carry out—can have the significance of indicating in it what also lies beyond (because it underlies) the possibility of our extending our concepts in just the ways we do. But the conception of the metaformal results as formally indicative in the way we have considered yields just such an account.[48]

4. Metaformal Realism, Ontological Realism, Temporal Realism

As we have seen, the realism that is recommended on either horn of Gödel's disjunctive conclusion has implications ranging far beyond those it has simply for "mathematical" truth. In particular, a similarly motivated metaformal realism is recommended wherever it is possible (or suggested on independent grounds) to be realist about *sense*. We obtain this realism as soon as we recognize that the *constitution* of sense in any particular domain is not simply the result of its *construction*, whether by means of the capacities or activities of a finite "human" subject or indeed by the *institution* of finitely specifiable regular procedures, rules, or norms of "intersubjective" practice.

Dummett's framework has the salutary benefit of resituating the question of realism as a question of the determinacy of sense rather than as the old question of the constitution of objects in relation to our (subjective) ways of knowing about them. But in applying it and especially in arguing for antirealism, he tends to assume that sense *must* be "provided" by means of socially instituted practices, if it is to be provided at all. By contrast with this, though, metaformal realism is a realism about the "provision" or constitution of sense that separates it from any constructivism, whether of a subjectivist, social-pragmatist, intuitionist, or any finitist kind. From the current perspective, the problem of this constitution is not distinct from the problem of the accessibility of the infinite to thought itself. For (linguistic) sense is in itself infinite, if only for the reason that *knowing* or *understanding* the sense of a single term involves, in principle, knowing how to apply it in an infinite number of cases, or use it in an infinite number of possible sentences.

If, as I have argued, the complex of results running from Cantor, through Gödel, to Tarski, shows the irreducibility of this access to any finitely specifiable procedure, it also thus motivates a *realism about sense and its givenness* that outstrips any determination of this givenness in terms of (finitely specifiable) capacities, abilities, faculties, or practices. Commitment to this realism can, as usual, naturally be formulated as commitment to bivalence in the relevant domain. The presupposition of realism in this sense is arguably essential to the *demonstration* of results such as Gödel's own incompleteness theorems, and it is equally essential to each of the two horns of the disjunctive conclusion he drew in 1951.[49] It is also a necessary formal precondition for the demonstration of the metalogical paradoxes (such as Russell's paradox) that condition the post-Cantorian orientations. Even if these are indeed taken to show specific points of the violation of the law of *noncontradiction*, it is essential to their positive demonstration that bivalence (i.e., the claim that each metamathematical statement is *either* true or false [or indeed, perhaps both]) be maintained.[50] The present suggestion is just that this sort of realism should *also* be extended to the "ontological" statements which specify the relationship between "being" and "beings as a whole" as specifications of the ontological difference.

With this in view, *metaformal realism* can then be simply defined as realism, in Dummett's sense, with respect to statements asserted on the basis of formal reflection, including reflection on the conditions and structure of the givenness of sense or meaning. These statements are to be seen as including statements "about" being "itself" (i.e., those statements using "being" as a singular term), as well as those about the givenness and availability of beings as a whole, and those about the difference between beings and entities. As applied to the ontological problematic, this realism about the constitution of sense is also thereby an *ontological* realism: one capable, that is, of maintaining a "realist" attitude not only with respect to beings but also with respect to the ontological difference and to being itself. The significance of this realism is that it must be possible to maintain bivalence *even with respect to "ontological" discourse*: that is, even the statements which articulate the ultimate basis for sense must themselves be subject to bivalence, bearing determinate truth-values (including the possibility of being both true *and* false) independently of "our" ways of knowing them, accessing them, or rendering them "intelligible."[51]

The result is a formally founded "ontological" realism that allows us realistically to characterize the logical, formal, and ultimately *temporal* structure of the ontological difference. If there is, as Heidegger constantly argues, some at-first-obscure structural relationship between being and time, to be elucidated by means of ontological investigation, then it is plausible that bringing it out will require a metaformally motivated ontological realism of the sort I have discussed. Moreover, as I shall argue in the next chapter, metaformal realism can indeed underwrite a *temporal* realism that clarifies in formally indicative terms the structure of the time of the world as such. In particular, ontological realism becomes temporal realism if bivalence (including the possibility of true contradictions) is maintained with respect to various *temporal statements*, including those whose senses are seen as having *ontological* determinants.[52] These include, crucially, (i) statements involving temporal determinations about events prior to and independent of the existence of conscious subjectivity (such as, for instance, the description of an event of evolutionary or biological history taking place prior to the advent of humans); (ii) statements having an irreducibly "indexical" or indicative form with respect to the present moment or "now" (such as "It is 5:30 now") and (iii) statements reporting and describing temporal processes of *change* and *becoming*. In each of these cases, bivalence and realism can be jointly maintained and defended, as I shall argue, by means of a metaformal indication of the ontological situation underlying the provision of their sense. Furthermore, as I shall argue, this indication provides a clarified understanding of the formal-ontological structure of the presentation of the present, or of the ontological sense of "given" time. This structure is underwritten by ontological realism and plausibly provides an account of the structure of the time of the world as such, quite independently of any assumption of its basis in a constituting or temporally founding subject.

6

Sense, Time, and Paradox

Within the onto-theo-logical tradition and the history it determines, the form of *self-consistent* and *complete* presence is recurrently understood as the metaphysical basis of the *being* of *whatever is*. But the more basic logical, ontological, and metaformal determinants of *this* form are not thought or more deeply interrogated. As a consequence of this, the sense of things is itself understood, within onto-theo-logy, in a twofold way: both under the *temporal* condition of their sensory experienceability and under the (putatively) *nontemporal* or *atemporal* form of their comprehension in thought. In creating the dualistic division that thus results between the "real" world of temporal becoming and change and the "timeless" world of ideal contents, onto-theo-logy thereby obscures the more basic ontological relationship between sense and time themselves.

This more basic relationship should, on the other hand, become available through a formally guided investigation of the ontological conditions of sense and presence. For if (as we have done since chapter 1) *sense* is defined in general terms as the *mode of presentation* of whatever is, then it is also necessary to recognize *time* as the unitary form of the *presentation* of *presence*, as (that is) the form of the being-present of whatever is thinkable, knowable, experienceable, or given, for or to a being that is itself basically conditioned and structured by time and the presence of a present "now." The investigation of ontological *sense* thus ultimately drives toward a clarification of the *unitary* ontological structure it shares with what may at first seem the distinct topic of *time*, the structure of the becoming-present of the present as such. It is by indicating this original structure, as I shall argue, that an attitude of metaformal realism can also support an ontologically articulated realism *about* time as it is "given" and available in general. This would be a realist doctrine of the constitution and structure of infinite time as it is available *both* to experience and to thought, both as recurrently and constantly "there" in an irreducibly situated "now" *and* as represented generally in thought, and thus countable or calculable *in general*.

The argument of this chapter is that such an ontologically basic structure of real time can be extracted from the onto-theo-logical tradition, once this tradition is itself subjected to critique on metaformal grounds. Specifically, the essential formal elements of a paradoxical and real structure of given time, as it might be thought "outside" the closure of onto-theo-logy, are already essentially at hand in *Kant's* (idealist and constructivist) problematic account of the "givenness" of time in his doctrine of the schematization of phenomena. This structure can be extracted from Kant's account, and separated from constructivism, once that account is critically liberated from two kinds of assumptions that still guide and constrain it. The first of these is the idea of a constitutive *subject* of faculties whose self-constituting and auto-affective activity ultimately gives time. The second is the idea of time's ongoing generation in the form of a *rule* that repeats itself self-identically ad infinitum. In developing both lines of critique, I partially follow Heidegger's own critical interpretation of Kant in *Kant and the Problem of Metaphysics,* bringing out the ontological basis of Kant's conception of the structure of finitude and its relation to the infinite in thought and interrogating it on metaformal grounds. With this critique, as I shall ague, the metaphysical duality that opposes the temporal finitude of the sensory to the eternity of the intelligible can cede to a metaformally and ontologically deeper structure of *originally given time.* This structure is indicated on the basis of a consideration of the form of time's reflexive self-givenness, which I shall describe and try briefly to characterize.

1. Kant and the Structure of Finitude

How is the ontological relationship between sense and time to be thought, in terms of the specific question of how the structure of "finite" thought relates to the infinite? In *Kant and the Problem of Metaphysics,* published in 1929 but drawing centrally on the interpretation of Kant already worked out in the "Logic: The Question of Truth" lecture course of 1925–26, Heidegger specifies *one* way in which a constitutive conception of "human" finitude may provide the foundation for an overall understanding of the total sense of entities in the world. On Heidegger's reading, Kant's program in laying out the grounding for any possible metaphysics in the *Critique of Pure Reason* depends, at its core, on a conception of human pure reason as *essentially* finite. Kant understands this finitude in terms of the requirement for human thought to grasp its objects by means of an *intuition,* which is, for him, the specific form of representation whereby knowledge is related *directly* to an *individual* object.[1] The essential finitude of specifically human thought is illustrated, for Kant, by the contrast between this knowledge and the possible knowledge of a divine or absolute intellect. Such an intellect would be capable, according to Kant, of an *intellectual* kind of intuition that would actually *create* the intuited object. By contrast, for Kant as Heidegger reads him, although human knowledge is always a synthesis of intuition and conceptual understanding, it is characteristically *finite* in that it stands under the necessity of *representing* objects which it cannot produce by itself and which therefore must be given to it from elsewhere.[2] For a divine intellect that did not stand under this necessity, it would also not be necessary to think (but only to intuit) in order to have all of its objects adequately. But since human cognition is thus constrained, even in thought, its characteristic finitude applies not only to the "faculty" of intuition but also to *conceptual* thinking by means of what Kant characterizes as the understanding and indeed, in a more basic sense, to the unity of the two.[3] The basis of this unity for Kant becomes more clearly visible, according to Heidegger, in considering the way in which this conception of essential finitude determines the Kantian idea of the *transcendental*. Specifically, Kant characterizes objects of knowledge in a twofold sense, both as *appearances* [*Erscheinungen*], insofar as they are knowable to us, and also as what the appearances are appearances *of*, that is, unknowable things in themselves. Since appearances, by contrast with things in themselves, stand under the specific limitative conditions of the forms of intuition, their nature as appearances is determined by the essential finitude of human intuition.[4] This is again to be contrasted with the infinite knowing that would be capable of knowing its objects *as* things in themselves. This contrast is, Heidegger suggests, the actual meaning of Kant's general claim that things in themselves lie "behind" appearances. The claim here is not that there are two types of possible objects of finite intuition, one lying behind the other, but rather that knowledge of beings *as* they are in themselves is essentially closed to human beings, as essentially finite knowers.[5]

For the same reason, according to Heidegger, it is necessary that human knowledge always involves an element of receptivity and that this element be *sensory* in that it operates through organs of sensation that allow (what Heidegger calls) the essentially finite Dasein to be affected by particular external objects.[6] Nevertheless, genuine knowledge is not simply knowledge that immediately represents what is present to one. Rather, it must also be able to make what is revealed "as revealed accessible with respect to both what and how it is for everyone at all times."[7] Kant understands this requirement of "representing in general" as fulfilled insofar as knowledge involves not only intuition but also "representation in concepts."[8] This determinative representing of something is an "assertion of something about something" or a predication.[9] The basis of this possibility of determinative representing, however, is for Kant the "faculty of judging [*Vermögen zu urteilen*]," or what Kant terms the "understanding." In the actual act of determinative judgment, the faculties of intuition and understanding are united in a synthesis and thereby "mediately" related to the determined object.[10] This synthesis "accomplishes" the making-evident [*Offenbarmachen*] of this particular entity as an object, making it available "in truth."[11]

This raises the question of the nature of the specific kind of synthesis between intuition and understanding that must occur for (true or false) judgment in general to be possible. The problem of the basis of this synthesis is in fact, Heidegger suggests, the deepest problem of the whole project of the *Critique of Pure Reason* in its attempt to lay a critical ground for any possible systematic metaphysics. Kant's attempt to solve it leads him to envision a mediating "common root" in the soul for both intuition and thinking; Kant characterizes this "common root" as a "power of pure imagination" that operates as a "blind though indispensable function of the soul."[12] This power is actually at the basis, Kant says, of all cognition whatsoever, and thus acts as the *general* function underlying all possible representation, whether intuitive or intellectual.[13] This includes what Kant describes in the "Transcendental Deduction" as the necessary condition for all possible objective representation, namely the transcendental unity of apperception. This unity has the form of a standing capability to unify intuition and the understanding. It is expressed by the "I think," which must, according to Kant, be able to accompany all of my representations.[14] But this capability itself remains dependent on the deeper "productive" synthesis of the imagination, which must accordingly function in a unitary way with respect to both sensory and conceptual operations of thought. This points, according to Heidegger, to a deeper presupposed or implied unification of the *totality* of the "universe of all beings" or of beings as such. This totality would be, on the one hand, the same totality that is available to the (infinite) divine intellect, endowed with the power of an originally creative intuition of things in themselves. But on the other, it would be constitutively subject to the power of an essentially finite thinking grounded in the imaginative synthesis that yields the transcendental unity of apperception, for us:

> The representing of unity, as pure thinking, necessarily has the character of the "I think." The pure concept, as consciousness of unity in general, is necessarily pure self-consciousness. This pure consciousness of unity is not just occasionally and factically carried out, but rather it must always be possible. It is essentially an "I am able." . . .
>
> Now what is represented in the unity which is held before itself in this way [*der durch sie vorgehaltenen Einheit*]? Perhaps it is simultaneously the universe of all beings [*das All des Seienden*], in the sense of the *totum*

> *simul*, which the *intuitus originarius* intuits? But this pure thinking is certainly finite, and as such it cannot from itself, through its representing, set the being in opposition to itself [*das Seiende sich gegenübersetzen*], not to mention simultaneously setting everything in its unity [*geschweige den alles zumal in seiner Einheit*]. The represented unity first awaits the encountered being [*das begegnende Seiende*]; and as such awaiting, it makes possible the encountering of objects which show themselves with one another. As non-ontic, this unity bears [*trägt als nicht-ontische*] the essential tendency of a unifying of that which is not yet unified in itself. That is why, following the clarification of Transcendental Apperception, Kant says of the unity which is represented in it: it "presupposes a synthesis however, or includes one."[15]

This deeper, presupposed synthesis is, however, just the "pure synthesis of the imagination," which is, according to Kant, the "ground of the possibility of all cognition [*Erkenntnis*]" as its necessary *a priori* condition.[16]

As Heidegger points out, this conception of the productive power of the imagination as the basic *a priori* condition for the possibility of any synthetic unification itself presupposes the givenness to intuition of *time*. As the unitary form of both inner and outer sense, time in its givenness as a form of ordering also conditions, according to Kant, all "modifications of the mind."[17] In particular, as Kant explains in the "Schematism" chapter, the pure power of the imagination is the capacity to link intuition and the understanding by forming an image or schema which is the image of a horizon within which particular objects can be encountered and experienced.[18] As such a "forming" of the horizon which also gives rise to the possibility of its being seen in a unified "look," the schematism also makes visible, according to Heidegger, "the ground for the possibility of transcendence." But since "transcendence is, as it were, finitude itself," this is also a "making-sensible" of the basic structure by means of which a finite creature can intuit anything at all.[19]

In invoking the "transcendental schema" as a "mediating representation," between the category and the sensible intuition, which is nevertheless homogenous to both, Kant invokes a "third thing" between them which is officially both intellectual *and* sensible.[20] Thus understood, the schema is the "formal and pure condition of the sensibility, to which the use of the concept of the understanding is restricted."[21] At the same time, the schema of a particular concept as the "representation of a general procedure" for giving it an image is what allows the concept, which is itself a rule, concretely to be applied to its various instances.[22] For example, in the case in which I think a "number in general" for which I do not provide a direct image, my thought is not itself an image or directly related to one but is rather the "representation of a method for representing a multitude [i.e., one having that number] in an image in accordance with" the concept in question.[23] In the case of empirical or mathematical concepts, the schema thus operates as "a rule for the determination of our intuition" in accordance with which an image is provided for a concept.[24] By contrast, in the case of *pure* concepts or categories (such as those of unity and plurality, or cause and effect, or substance and accident) there is and can be no such image or (accordingly) any "method" for providing one. Here the schema is thus simply "the pure synthesis, in accord with a rule of unity according to concepts in general, which the category expresses . . ."[25] In either case, however, the schema is, as Heidegger glosses it, a "representing of the rule" that the concept is, which in a certain way "bring[s] the rule into the sphere of possible intuitibility."[26]

In this way, the schemata allow for the possibility of the basic connection between the intuition and the understanding by presenting or representing concepts in such a way that they become intuitable. Such presentation itself manifests the basic structure of transcendence, which consists in the finitude whereby human thought and intuition are jointly conditioned by the necessity of representing something exterior to themselves. But the schemata of pure concepts are explicitly themselves nothing other than pure, *a priori*, and transcendental determinations of time.[27] Heidegger suggests, in particular, that the schema of the category of *substance*, as the category signifying that which provides the ultimate ground of all persistence, is, for Kant, the most basic "pure image of time [*Ihr Schema muß die Vorstellung des Zugrundeliegens sein, sofern es sich im reinen Bilde der Zeit darstellt*]."[28] As Heidegger interprets it, this is the image of a constantly successive sequence of "nows" which is also *permanent* and eternally *persisting* in the sense of never running out. In this "pure sequence of nows" that "now time [*Jetztfolge*]" represents, time is always "now [*Nun ist die Zeit als reine Jetztfolge jederzeit jetzt*]."[29] As the constant substrate of this constantly "flowing" sequence, time as schematized in the schema of substance is that which endures, giving "the pure look of something like lasting in general."[30] In this way, through the schematism that thus renders the "pure image of persistence" visible, "a being which as such is unalterable in the change can show itself for experience [*so daß . . . für die Erfahrung ein im Wechsel unveränderliches Seiendes also solches zeigen kann*]."[31] This schematization of time as such thus functions, for Kant as Heidegger interprets him, as "the ground for the inner possibility of ontological knowledge."[32] It does so by giving to experience a "preliminary enclosedness to the horizon of transcendence."[33] In this *a priori* giving of time as a “unique, pure, universal image,” it thus gives an image to the “single and pure ontological horizon” which is the condition for the possibility that any “being given within it can have this or that particular, revealed, indeed ontic horizon.”[34] As such a “pure self-giving,” it makes visible to a finite creature the ontological structure of its own finite transcendence.

This structure of finite transcendence essentially involves that a subject thus constituted can be affected by something outside itself, but also in such a way that it “bears and makes possible” the general concept or category.[35] This possibility of being affected from without but in such a way as to facilitate the concept is a general one. As such it must also characterize, according to Kant, the “formal conditions” of the way in which we represent all temporal relations of succession, coexistence, and endurance. For Kant, however, these “formal conditions,” being purely relational, do not and cannot represent any thing in itself but must instead represent things only insofar as they are “posited in the mind.”[36] This is what leads Kant to consider the basic temporal form of inner sense as “nothing other than the way in which the mind is affected by its own activity, namely this positing of its representation,” or, as Heidegger puts it, as the mind’s “pure self-affection.”[37] As Kant further suggests, this self-affection itself conditions the possibility of the apperceptive consciousness of self, whereby the self appears to itself *as* it is affected by itself.[38] Thus the ultimate significance of human finitude for Kant, according to Heidegger, is not to be seen simply in the fact that the finite intellect is necessarily affected from without, in sensation, by something other than it. It is rather, more deeply, that this possibility of affection from without is itself dependent on a pure *self*-affection which, in yielding the form of time, first constitutes the apperceptive unity of the self to begin with.[39]

But, according to Heidegger, although Kant thus sees the way in which the givenness of time as a kind of universal self-givenness conditions all possibility of representation and thus of objects of experience, he nevertheless understands this givenness in a way that is ultimately *paradoxical* or *contradictory*. This is because he understands this givenness of time *both* as the result of an *active* and “productive” capacity of synthesis *and* as shaped by a formal condition with respect to which it is *passive* and receptive. As the agency ultimately capable of uniting the faculties of the intuition and the understanding, the transcendental power of imagination must also unite receptivity and spontaneity in an original and noncomposite way.[40] Accordingly, since the transcendental imagination is the ultimate basis for the givenness of time, time itself must be given, paradoxically, both receptively “from without” and spontaneously “from within.”[41]

Insofar as both operate as subsistent and perduring conditions for all possible representation, Heidegger suggests, time and the "I think" of transcendental apperception are, for Kant, ultimately the same.[42] But if the "I" of the "I think" gives itself time through the original structure of a pure self-affection that is irreducibly both spontaneous and receptive, this means that it is also itself first constituted by this very giving. It thereby becomes possible, according to Heidegger, to challenge Kant's official view, according to which neither the "I" nor time itself are in fact "in time." Indeed, without simply denying this official view, it here becomes possible to ask whether Kant's attribution of permanence both to the "I" as transcendental subject, and to time as the constant succession of "nows," might in fact point to a deeper way in which both are temporal. Heidegger puts this as the possibility that the "I," far from being simply atemporal, is "so 'temporal' that it is time itself," and indeed that it only becomes possible *as* "time itself, according to its ownmost essence."[43]

If the "I" of transcendental apperception which forms the basic unity of thought must be thus be placed in an essential relation to the pure self-affection that is also the basic form of time, it is also necessary, according to Heidegger, to rethink the temporal status of the constitutive forms and rules of pure thought itself. This is so, according to Heidegger, even with respect to what Kant sees as the "highest principle of all analytical judgments," namely the principle of noncontradiction. Heidegger notes that in introducing the principle, just after the schematism chapter, as the basic principle underlying all analytic knowledge and a sine qua non of all knowledge whatsoever (whether analytic or synthetic), Kant emphasizes that it is a "merely logical one" and therefore must not be understood as limiting its claims to those involving relationships of time.[44] Thus, it is necessary, according to Kant, to replace what he cites as the traditional formulation of the principle, namely, "It is impossible for something to be and not to be at the same time," with an alternative formulation making no mention of time at all.[45] If, however, as Heidegger suggests, the basic structure of the "I think" of apperception must be reconsidered in its relationship to the underlying structure of self-affection at the basis of time, then even Kant's attempt to interpret the principle of noncontradiction as extratemporal in this sense must also be rethought on this basis.

Insofar as it is possible to maintain that the "I think" of transcendental apperception remains simply outside time, it is also possible for Kant to find in it the underlying principle of unity and persistence that allows for the principle of noncontradiction to appear capable of applying to *all* objects as such, insofar as they are thinkable at all, without bringing them into any specific relation to time. Within the scope of Kant's assumption of the secondary, derived conception which places both the "I" and its thought outside the realm of the "within-time," the correction which consists in deleting the "at the same time" from the statement of the principle is thus justified. Nevertheless, by bringing into view the more original link between time and the basis of thought in the structure of self-affection, it is possible, Heidegger suggests, to interpret this "at the same time" in a wholly different way. Here, in particular, though it is no longer simply a matter of the copresence of beings within time, it may nevertheless be seen as involved in the actually *temporal* character of thought in a more basic and original way.[46] In this more basic sense, the possibility of recognition at the basis of any possible judgment of identity does not depend on the ontic copresence of beings in the "at the same time" of a present moment. Accordingly, the formulation of the principle of noncontradiction does not have to exclude such a contradictory simultaneous copresence of ontic properties. Rather, according to Heidegger, it points back to the "belonging together" of thought and intuition, or of spontaneity and receptivity, in the "unity of the same essence," one which is predicated on the original constitutive structure of temporal self-affection that is also the original form of the givenness of time itself.

2. Temporality and the Grammatical Investigation

What is the form of a hermeneutic investigation into the ultimate formal/structural conditions for the givenness of time in its essential nature, and by what means can such an investigation proceed? At *Philosophical Investigations* (hereafter *PI*) 89, just after invoking Augustine's famous puzzle about the essence of time (namely, that he seems to know what it is when nobody asks, but when he is asked, he does not know), Wittgenstein specifies one such method as the "calling to mind [*Besinnung*]" of what we in some sense already have lying before us, "open in plain view":

> Something that one knows when nobody asks one, but no longer knows when one is asked to explain it, is something that has to be *called to mind* [*etwas, worauf man sich* besinnen *muß*]. (And it is obviously something which, for some reason, it is difficult to call to mind [*etwas, worauf man sich . . . schwer besinnt*].)
>
> 90. We feel as if we had to *see right into* [*durchschauen*] phenomena: yet our investigation is directed not towards *phenomena*, but rather, as one might say, towards the "*possibilities*" of phenomena. What that

> means is that we call to mind [*wir besinnen uns*] the *kinds of statement* that we make about phenomena. So too, Augustine calls to mind [*besinnt sich auch*] the different statements that are made about the duration of events, about their being past, present, or future. (These are, of course, not *philosophical* statements about time, the past, the present and the future.)
>
> Our inquiry is therefore a grammatical one.

What is at issue in the context is the particular "depth" that appears to characterize logic as something "sublime," something that must be fixed in advance of any empirical investigation and must thereby be seen as capable of determining the possibilities of phenomena always already in advance. This character of "depth" has long been seen, in temporal terms or ones analogous to them, as that of the "*a priori*." The problem of the way in which it has been given to us, or the way it can be explicitly retrieved, is thus nothing other than the temporal problem of the *a proricity* of the *a priori* as such. The problem of the "knowledge" of the essence of time which is given to us as finite beings who themselves live and define the distinctive "possibilities" of their life in terms of the way time is given to them has itself long been thought (and is thought, also, by Augustine) as the problem of the "accessibility" of this *a priori within* this empirical time of factical life and knowledge. Given the general form of this problem, Wittgenstein's choice of Augustine's question *about* the nature of time as an example that illustrates it is thus (though others might perhaps have been chosen) by no means adventitious. As the sort of question that one feels one can answer when nobody asks, but is unable to answer when it is asked, the form of Augustine's question is itself indicative of the problematic of the nature of given time which is its theme. This is the problematic of the recovery of the original structure of logical and temporal possibility from within the temporal life of a finite being conditioned by it.

In response to this linked problem of logic and time, Wittgenstein here suggests that clarity can result from the recollection or "calling to mind [*Besinnung*]" of what we (in some sense) already know. This involves recalling the kinds of *statements* we make ordinarily—statements, for example, "about the duration of events [and] about their being past, present, or future."[47] The suggestion invokes Augustine's own procedure, but it may also be seen as evoking or resembling one of the original temporal figures of the specific character of rational knowledge in the Western tradition, namely Plato's invocation of *anamnesis* as the recollection, under the conditions of an embodied life, of what one already knows but has in some way, due to this embodiment, necessarily forgotten. The point of connection is not in any assumption that Wittgenstein shares with Plato about the necessary opposition of the sensible and embodied over against the supersensible and atemporal in itself, or indeed in any mystified figure of memory or recollection *beyond* the temporal boundaries of an embodied life. Rather, it lies in what is here suggested about the original *temporal* structure of a critical inquiry into the foundations of the "prior" givenness of sense that have long figured, in the tradition, as the *a priori* order of possibilities fixed always already in advance.

The *Philosophical Investigations* develops this inquiry largely through two interlinked movements of critique. The first, the so-called "rule following considerations," takes up the question of what determines the *application* of a rule across the infinite number of instances of its possible use. It culminates in the "paradox" of *PI* 201, according to which "no course of action [can] be determined by a rule, because every course of action can be brought into accord with the rule" on some interpretation of it. The second, what has been called the "private language" argument, challenges the idea of a privileged interiority in which it is possible for one to present "private" items of experience, sensation, or thought to oneself, quite independently of public and essentially diachronic usage. The critical targets of the two strands of critique can be seen as converging on a unitary idea, with relevance to Kant's attempt to account for the generation of time through the schematism. This is the idea of meaning as constituted through the interior activity of a representing subject able to guarantee the correctness of her own usage by recurrently presenting to herself a finite item—for instance, an image or picture—which remains in each case identically and evidently the same.[48]

Kant's idea of the transcendental schematism is, as we have seen, the idea of a *power* of the subject that unites the faculties of the understanding and the intuition, and which is thereby the common root of the spontaneity characteristic of the former and the receptivity characteristic of the latter. The power is, specifically, that of applying a *concept* or *category* (once attained) to the *indefinite* or *infinite* number of appearances that can correctly be subsumed under it. What accomplishes this power, in the subject, is explicitly, for Kant, the "representation of a rule" that connects the concept, as representation, to this indefinite number of appearances. In the case of the schematism of substance—wherein, as we have seen, Heidegger locates Kant's most general account of the form of the subjective givenness of time—this representation is that of a pure rule of indefinite persistence which must apparently be presupposed as underlying the unified temporal form of inner as well as outer sense.

To see how Wittgenstein's critique of rule-following itself points to the deeper paradoxical structure underlying Kant's account, it is helpful to consider the relationship between the idea of a rule which he most directly criticizes—that of a "rail laid to infinity"—and the constitutive idea of the self-identity of a rule as an *infinite* repetition of the same. At *PI* 213, an interlocutory voice considers the possibility that the ability to continue a series indefinitely beyond its initial segment consists in having chosen *one* interpretation of it. In view of the "paradox" of interpretation of *PI* 201, *this* suggestion is untenable: the explanation in terms of the initial adoption of *one* interpretation capable of eliminating all doubt is here neither necessary (since the possibility of doubt to which it responds in fact *need* not arise) nor sufficient (since the provision of an interpretation in this sense *cannot* "fix" the infinite number of cases uniquely, all by itself). This leads the interlocutory voice to the suggestion that "intuition" is needed to remove the (possible) doubt in each case:

> Only intuition [*Intuition*] could have removed this doubt?—If intuition is an inner voice—how do I know *how* I am to follow it? And how do I know that it doesn't mislead me? For if it can guide me right, it can also guide me wrong.
>
> ((Intuition an unnecessary evasion.))
>
> 214. If an intuition is necessary for continuing the series 1 2 3 4 . . . , then also for continuing the series 2 2 2 2 . . .

The conception that underlies the thought that the continuation of a series, if it cannot be determined once and for all in advance by a single "interpretation," requires a new intuition at each instance would also, if tenable, have to apply to the seemingly most basic kind of rule, the one that involves only the repetition of the same ad infinitum. And if intuition *here* were to function as a kind of inner voice, then it would apparently be possible here *too* that it might mislead. To block this conclusion, the interlocutor now envisages a kind of absolute givenness of identity in the form of the self-identity of the same (*gleich*) in its indefinite repetition:

> 215. But isn't at least the same *the same* [*Aber ist nicht wenigstens gleich*: gleich]?
>
> For identity we seem to have an infallible paradigm: namely, in the identity of a thing with itself [*Für die Gleichheit scheinen wir ein unfehlbares Paradigma zu haben in der Gleichheit eines Dinges mit sich selbst*]. I feel like saying: "Here at any rate there can't be different interpretations. If someone sees a thing, he sees identity too [*Wenn er ein Ding vor sich sieht, so sieht er auch Gleichheit*]."

Wittgenstein's critical response comes swiftly:

> Then are two things the same when they are what *one* thing is [*Also sind zwei Dinge gleich, wenn sie so sind, wie* ein *Ding*]? And how am I to apply what the one thing shows me to the case of two things?
>
> 216. "A thing is identical with itself [*Ein Ding ist mit sich selbst identisch*]."—There is no finer example of a useless sentence, which nevertheless is connected with a certain play of the imagination [*der aber doch mit einem Spiel der Vorstellung verbunden ist*]. It is as if in our imagination we put a thing into its own shape and saw that it fitted.

If the idea of a thing's self-identity is empty in this way, then it cannot be used positively to undergird the conception of the underlying possibility of the repetition of a rule according to which this repetition is itself that of a self-similar item ad infinitum. Neither can, for the same reason, the assumption of the iterated intuition of a directive, or indeed, and again for the same reason, of a "decision" made anew in each case.[49] In the Kantian jargon, neither the receptivity of a subject in the pure intuition that it gives to itself in submitting itself to the rule, nor its spontaneity in the "decision" that would consist in each case, in the self-giving, can ultimately account for what it is to be "able" follow the rule "correctly" or to know, grasp, or understand it. For neither the spontaneous self-giving of the rule in finite form at each step, nor one's receptive relation to it (assumed as given) can succeed in overcoming the radical aporia of finitude and infinitude that stands at the center of the very conception of rules as regular structures of the infinite repetition of the same. Failing in the attempt to account for the development of the rule in either of these ways, the conception finally attempts to base itself in the idea of an "infallible" and absolute paradigm or criterion for this infinite repetition in self-identity. Here it is indeed tempting to seek a ground for the purported substance of this criterion in an operation of the imagination, the "play" by which it doubles the (single) object and finds it once more "identical with itself." But this final attempt, too, fails in that the self-identity of the same does not and cannot provide a *paradigm* for anything; all it can do is indicate a more original question about the basis of identity itself.[50]

If, then, any account of the actual basis of the "infinite" understanding of how to go on in terms of the provision to consciousness (whether from within or without) of any finite item must fail (on pain of Wittgenstein's paradox at *PI* 201), then Kant's schematism as the representation of a rule for subsumption must apparently fail for the same reason. For if it is the finite representation of the rule that is supposed to account for how a concept can be applied in an indefinite number of cases, then there inevitably arises the question, which Wittgenstein pushes, whether this rule itself needs to be interpreted (by a further rule) in order to be applied correctly in each case.[51] Of course, Kant does not raise the question of whether the schematism of a concept can be *mis*applied, or applied in a misleading or incorrect way. But it is the effect of his way of considering the possibility of subsumption to be given to the subject—namely by means of the provision of the "representation of the rule" which the schematism is—that this question is unavoidable. Once we pose it, along with Wittgenstein, it is no longer possible simply to assume the guaranteed correspondence of the "I can" of subjective schematization with the order of things as they are (even if this be, explicitly, the order of things as appearances and not things in themselves).

3. Paradoxical Structure of Given Time

If Heidegger is right in his reading of Kant, it cannot be the case that time is produced through the constitutive capacities of a representational subject. But it is nevertheless possible, as Heidegger himself suggests, that the paradoxical structure discernible in Kant's picture may point, if separated from some of Kant's own assumptions, to a more original and basic connection of being and time themselves. In particular, it may be possible to develop the implications of this structure of temporal auto-affection even *outside* the context of the assumption of a rooting of time in the activity of the subject or in its ability to represent a rule. As grounded in a formally motivated realism, the illumination would then shed light on the basic *ontic-ontological* structure whereby there "is" something like time at all. Here, we are accordingly looking for this structure as the formally underlying basis for the possibility that anything like time is *given* at all—both in the empirical experience that presents it in immediate presence, and in the "intellectual" form of its abstract availability to thought, as such and "in general."

What, then, remains of Kant's picture of the original givenness of time, if we purge it of the assumption of its rootedness in the capacities of a representing subject but attempt nevertheless to retain its definitive connection to the finite/infinite problematic it involves? If we drop the constitutive assumptions of Kant's idea of the representational subject as conditioning all possible objectivity, we cannot retain the figure of a subjective *activity* of auto-affection whereby the subject both gives itself time from itself and is first constituted by the very form of its being given from outside. Still, in connection also with Heidegger's own account of "authentic" temporality in *Being and Time,* it may be possible formally to retain and characterize, as nevertheless capturing the formal and ontological core of Kant's account, an underlying structure that we can also see as forming the real basis of given time. It can be articulated by considering the constitutive structural features of **reflexivity** (or self-relation), **totality**, and **paradox**.[52]

1. **Reflexivity**: before and beyond the contradictory structure of activity and passivity that Heidegger locates in Kant's account of time as pure self-affection, he suggests a more basic structure of given time in which time temporalizes "itself." In *Being and Time*, this autotemporalization is discussed as time's "giving" itself in the three ecstases, thereby yielding what can be grasped as a past, present, and future, "across" and on the basis of the ontological difference.[53] In this, the structure of a basic reflexivity, or self-givenness, is retained in the ultimate and basic structure of given time. This reflexivity is not to be understood as if it were the representation of a self to itself or as if it depended on a subject's self-reflection in thought or consciousness. It is, rather to be understood as the formal/structural (and ultimately ontic-ontological) condition for any such self-relation as it empirically or factically can exist.
2. **Totality**: Although Kant, according to Heidegger, understands the underlying structure of given time ultimately in terms of what he (Heidegger) calls "world-time"—what is for Heidegger the merely extant time of the pure succession of "nows"—Kant's analysis stops short, according to Heidegger, of considering the actual structural relationship of given time to the phenomenon of Dasein's "transcendence toward world" and thereby its relation to the phenomenon of world itself.[54] In and through this relation, Dasein, as ontic-ontological, is able to transcend itself *toward* that which ultimately determines the ways of being of entities "as such and as a whole," or toward their being. As soon as this structure is brought into view, the question arises of the form of the relationship between time's self-givenness and the constitution and ontological provenance of the *totality of entities* as such.
3. **Paradox**: Through the combination of the first two features, the givenness of time—which must be the basis for the time-determinability of entities as such and in general—is also always irreducibly *at* or *as a* time, located as the phenomenon of *a* "present" moment or in the form of presence. This means that the ultimately constituting agency must, if the "intratemporal" is grasped simply as an ontic field, appear to be at once *inside* and *outside* this field. It is then capable, through (what must then appear as) its paradoxical agency, of constituting *all* of time in general, as if from "outside," but only

from a position irreducibly located within it. This is the underlying structure of the paradoxical aporia that Heidegger locates in Kant's conception of a constituting and constituted transcendental subjectivity, and it is irremediable as long as what is in view is the simply *ontic* existence of the constituting entity. But with the suggestion of an underlying *ontic-ontological* structure at its basis, the underlying paradox can be (if not eliminated) brought out in its real formal structure and viewed as a positive structural condition for the existence and the sense of (ontic) intratemporality itself.

With these essential features in view, we can return to the articulation of the four orientations of thought, this time understanding each in terms of how it proposes the structure of the givenness of time as such. As we saw (chapter 5 above), each of the four orientations involves a specific conception of the infinite, and each such conception also articulates a specific figure of the givenness of the world, or the ontic domain as such. These conceptions of the relationships of thought, or its located position, to the totality of the thinkable are, just as such, figures of the relationship or nonrelationship of finitude "to" a world in general. But as four structural figures of this relationship they are also therefore four figures of the givenness of the thinkable *time* of the world as such. Accordingly, they are also orientations that unfold the determinate configurations of the appearance, in what is here contemplated as the relationship of thought to beings in general, of what is itself without figure, the ontological difference between being and beings. They compose and decompose the conditions of any such figuration itself, according to the underlying constitutive ideas of consistency and completeness, as the original metalogical problematic of limits formally/structurally indicated in the temporal form of a "finite" life.

In particular: the *onto-theological* orientation determines the time of the world as given on the basis of the absolute time of a superbeing within which is lodged the power of an original creation, even ex nihilo; here, worldly time, whether limited or unlimited, is the time of an *ens creatum* secondary to and derivative of an original absolute eternity, outside or beyond it. With the *constructivist* orientation (as, exactly, in Kant) thinkable time is thought as the outcome of the determinate *activity* of a self-positing agency, capable in itself of unlimited potential continuation, though never given or even possible as a *determinate* and *completed* infinity.[55] With the *generic* orientation, time is thought as the periodization or gap between the concrete events of the transformation of finite conditions of thinkability which procedurally construct the basis of constituted sense, and the insistence of the "eternal" truths which draw them forth in accordance with the stringent law of procedural consistency. In the *paradoxico-critical* orientation, the infinite time of the world is originally given as the paradoxical structure of a reflexivity which is, with respect to the totality it involves, both a giving and a given, and therein stages the original structure of contradiction as limit-paradox.

From the position of the paradoxico-critical orientation, the structure of "world-time" as given can now further be specified in a twofold way, as consisting in two paradoxical structures of its givenness "across" the ontological difference. These structures articulate the paradoxical form of time as it is given both "from outside"—in relation to the time of the world *as a whole*—and "from inside"—in relation to the constantly "flowing" character of the *present* as such.[56] Both paradoxical structures evince the essential relation of change, motion, and becoming to negation and contradiction. As we shall see, as involving the essential features of totality and reflexivity (features 1 and 2 above) both have the form of *inclosure* contradictions, or contradictions at the limits of what can be thought or experienced. In the specific aspects of givenness that they articulate, they correspond to what is understood within the metaphysical tradition as, on the one hand, the givenness of world-time to thought, and on the other, its givenness to experience. However, as we have seen, in the present context the givenness of time can no longer be thought as dependent on a subjective structure of faculties, or as articulated essentially according to the duality of (temporal) experience and (atemporal) thought that organizes that structure for Kant. Instead, the twofold structure here provides an ontological schematization of the contradictory givenness of the time of the world, in general and as such.

3a. First paradoxical form of world-time: Cosmological paradox of the totality of time. Beginning with the present moment, it is possible for me to think the sequence of past times preceding it, or the sequence of future times following it, as unlimited in either (or both) directions. In either direction, the thought of a limit to the series—a first moment in the time of the world, or a last—immediately engenders the thought of the further continuation of the series. Although it is apparently possible to consider the totality of times, it is thus not possible to think this totality as bounded in the past or in the future without arbitrariness (that is, it is not possible to think it as bounded without denying the obvious continuability of the series). However, to consider such a totality is already to consider it as bounded.

This paradoxical structure of the given time of the world is essentially formulated by Kant in his consideration of the cosmological antinomies. The antinomies, on Kant's description, are contradictions of reason with itself with respect to the idea of *totality* involved in its ongoing search for the conditions of phenomena. They concern, respectively, the completeness of the whole of all appearances in space and time, their spatial and temporal divisibility, the origination of phenomena by means of natural causality or spontaneity, and the conditioning of all phenomena by an absolutely necessary being.[57] In the first antinomy (that of the whole of appearances), Kant considers specifically the question of whether the world has a beginning in time and whether it is bounded in space.[58] As Kant points out in the argument for the temporal version of the antithesis (i.e., the claim that the world has *no* beginning in time), the thought of a first moment in the temporal series already demands the thought of the moment that precedes this one and beyond. On the other hand, in the argument for the temporal thesis (i.e., that the world has a beginning in time) Kant argues that it is impossible for there to be no beginning of the world in time. Thus if the world indeed exists as a whole, it both has, and does not have, a first moment in time.

As Priest notes, the contradiction thus formulated can be understood as having the form of an inclosure paradox.[59] Here *Existence* is assured by the idea of a totality of moments (leading up to the present one). *Closure* is guaranteed by the thought, for any moment in world-time, that its predecessor is still a moment in world-time. And *Transcendence* is just the operation of proceeding in thought from one moment to its predecessor. An exactly parallel argument, moreover, can be run in the other direction, as bearing on the totality of future times and turning on the availability, for each moment, of its successor. Kant's own strategy, as with all of the antinomies, is to argue that the contradiction shows the falseness of the premise on each which of the contradictory claims is based: that there *exists* a world as a whole, or (in terms of the first antinomy) a completed series of events leading up to the present. According to the solution, although the world is *never* given as a (finite or infinite) *totality* of appearances, it is nevertheless always possible for reason to pursue the series of conditions for any given conditioned one step further: looking backward, in particular, it is always possible to seek the prior condition for a given phenomenon. But through this "successive synthesis" one never attains a total or complete representation of the world as a whole. For Kant, this further provides an argument for transcendental idealism, which involves the claim that there is no *world of appearances* as a whole: the (seeming) world of spatiotemporal appearances is always potentially extensible through the synthetic activity of the subject, but never exists as a complete whole in itself. As Priest notes, however, this strategy simply amounts to denying *Existence,* the claim of the (temporal) existence of the world as a whole. It is motivated, within the Kantian context, by what is essentially a strategy of *parameterization*: the reality of the world of appearances is denied, in view of the contradiction, by means of essential reference to a "higher" level of actual existence: that of the realm of things-in-themselves, seen as independent of space and time.[60] Given Kant's assumptions, the denial is necessary in order to preserve consistency on the level of the totality of what can be known; but its (significant) cost is that it is not impossible to maintain realism about the generation and being of time itself.

3b. Second paradoxical form of world-time: Kairological paradox of passage or of becoming in the now. The present comprehends all that is happening *now*—not only what is happening with, to, or *in me* but all processes, objects, and events that are presently extant or occurrent, anywhere in the world. But one of the things that is happening *now* is that the present moment is "already"—right now—passing away and becoming past, while (simultaneously) another moment that *was* futural is becoming present. The *present* moment is thus always already "becoming" another moment, and is thus never simply "itself." The "now" is, as such, thus both fully present and not fully present. Along with this, each process or phenomenon of something changing involves that something *is* the way that it is—and also that it *is not* that way, in that it is becoming something else.[61]

This paradox concerns the basic character of worldly *becoming* as such. As far as its statement and structure are concerned, it is indifferent whether it is stated in terms of the continuous *becoming* of things in the world or in terms of the continuous becoming (i.e., the becoming-past) of the "present" itself. (Indeed, since we are here taking the world to be the totality of *what is*, we can take these two forms as essentially equivalent.) Like the paradoxical cosmological form above, it also bears essentially the attributes of totality and reflexivity, but it bears them (this time) in a different way. Here the relevant totality is not the totality of all times in the past or future (or both) but the totality of all that is happening *now*—that is, all the processes, events, and entities (anywhere in the world) that are "presently" occurrent.[62] Reflexivity is essentially involved in that the "now" must itself be seen as essentially becoming past and future, that is, becoming *other* than *itself*. But one of the things happening *now* is this very becoming. Indeed, given these features, it is possible to put this paradox as well in the form of an inclosure paradox. Here, *Existence* is the existence of the totality of all that is happening *now*. *Transcendence* is guaranteed by becoming—that is, by the fact that the "now," along with everything it comprehends, is (right now) becoming something other than now, becoming no longer (i.e., not) now. And *Closure* is guaranteed by the fact that this becoming is itself among the things that are happening *now*. Given this, contradiction results immediately: the present both is, and is not, identical with itself (because its becoming both is, and is not, happening in the present).[63]

As we have seen since chapter 1, the question of the possibility of temporal becoming is decisive at the origin of the logical (or rather onto-theo-logical) tradition inaugurated by Plato on the basis of his response to Parmenides. It is Parmenides's original posing of the problem of change and becoming in terms of just such a contradiction, and his consequent denial of the reality of temporal change, which produces, when combined with the Heraclitean idea of the ubiquitous temporal becoming of the sensible, the form of the logical solution that the Eleatic Visitor of the *Sophist* gives to the question of the reality of becoming and change. Kant's own solution, under the condition of transcendental subjectivity, refers what is there seen as the merely *apparent* becoming of things outside me to a real activity of transcendental synthesis occurring (as he supposes) inside me. But as we have seen, this parameterizing solution suffices to resolve the contradiction at the root of the givenness of the present only by reinscribing it as the paradox of the transcendental subject's own status, as both *inside* and *outside* the time it conditions, and both *passive* and *active* in the conditioning. And this approach not only is unmotivated in that it ultimately fails to achieve its ostensible aim—namely to resolve the contradiction involved in becoming as such—but moreover cannot serve as the basis for a realism about world-time. For even if it were successful in resolving the contradiction of how time is given for the *individual* subject (it is not), the Kantian solution would not account at all for the present as it is given *simultaneously* and universally for *everyone* in general.

If, by contrast, the underlying paradox of the temporal present is brought out and clarified in the way I have suggested, it can be seen as underlying a coherent (although paradoxical) realist account of the givenness of the universal time of the world. The decisive consideration here is that whether temporal becoming is referred to a source "within me" or "outside me," it is nevertheless itself a happening *within* the totality of what *is* happening now. To construe this happening realistically is just to construe it among the things that are happening *within* the world; and to refuse to parameterize from the (ultimately incoherent) perspective of an "outside" of time from which temporality is constituted just is to affirm the basic contradiction that thereby results. Both the Parmenidean and the Kantian denials of the original paradox amount to varieties of this strategy of parameterization, and both thereby yield the claim that the passage of time is ultimately (or, for Kant, transcendentally) unreal.[64] By contrast, the two assumptions that (i) the passage of time is irreducibly real rather than illusory and (ii) time and its passage are always given to something that itself is situated *within* time are jointly sufficient to yield the basic and irreducible contradiction of world-time sketched here.

This contradiction is itself closely linked to the problem in terms of which much twentieth-century and contemporary discussion of the question of the reality of time has been couched. This is the problem posed by McTaggart of the relationship and reality of what he terms the "A-series" and the "B-series."[65] The problem, as McTaggart states it, is briefly this. We discuss events and positions in time by placing them in either of two fundamentally different kinds of series. In placing a temporal position in what McTaggart calls the A-series, we situate it as past, present, or future (relative to the present moment): events have either taken place, or are taking place now, or will take place. This determination is not permanent, since something that is now located in the future will (later on) be present and then (still later) finally be past. On the other hand, by placing an event in the B-series, we give it a determination relative to other events simply as earlier than, later than, or simultaneous with them. The second series, by contrast with the first, is permanent: the temporal relations it ascribes are conceived as fixed and unchanging. The problem that McTaggart points out, in order to argue for the ultimate unreality of time-relations (*tout court*), is that to place events or positions in the A-series is to affirm contradictions. For pastness, presence, and futurity are contradictory determinations, each of which excludes the other two; but placement of any event or position in the A-series involves ascribing (at different times) each of the three determinations to it. Nor can the contradiction be removed by maintaining that the determinations are not contradictory, since what is presently in the future *will be* in the present and then *later* will be in the past. For this description simply produces an iteration of the same structure, which remains contradictory. If, however, the A-series is deemed unreal because of the contradiction, it is then unclear *what* is related in the B-series itself, and so it too must apparently be rejected as unreal.

McTaggart's own idealist solution is to reject the reality of *all* temporal relations: nothing is in time, because nothing *actually* stands in the relations constitutive of either the A- or the B-series. On this suggestion, there remains the problem of explaining the *appearance* of time and change, even if it is illusory. Partly for this reason, more recent discussions of McTaggart's problem have taken the form of either attacking, or defending, the reality of *change* or *becoming* which is involved in the assumption of the reality of the A-series. By contrast with McTaggart himself, though, even those in the recent discussions who have denied the reality of the A-series have typically nevertheless defended the reality of the B-series, holding that it alone is to be accorded (temporal) reality.

In this defense, the claims of such "B-series theorists" have been bolstered by their appeal to a theory of the linguistic functioning of temporal indexical terms in language, such as "now."[66] On this account, such terms are indexical (or perhaps "token-reflexive") in that they have a meaning which is itself variable and determined by the *context* of the utterance of the token term. Thus, a token of "now" functions, in ordinary discourse, to pick out the time at the moment of its utterance, analogously to the way "here" functions to pick out the immediate location of its utterance, or "I" to pick out the speaker. The illusion of the existence of change in the sense of the A-series, on this kind of account, is to be accorded to the mistake of thinking that temporal indexicals have any *further* meaning, beyond that already included in this indexical functioning relative to contexts. Just as the world could be given a full spatial description without mentioning what location "here" is, it would be possible (according to these B-theorists) to give a *complete* description of the world without mentioning any of the seeming facts that are ascribed by placing events and temporal locations in the A-series. Instead, temporal indexicals are treated as, at most, pointing to relational facts about the temporal locations of various (token) utterances, all of which can readily be accommodated within the B-series alone.

For the B-theorist, there is no sense in which things actually change from being one way to being another, or in which things become different from the ways they were, other than that in which there are different states of things at different temporal positions or locations. Accordingly, many A-series theorists have held that the B-theory leaves out what is (for them) an evident and indisputable fact, that of the reality of temporal change or becoming, and have often appealed to a basis for this self-evidence in subjective intuition or given experience. By contrast with this line of defense, however, from the present perspective the reality of becoming and change can be defended on a different kind of ground, one that makes clearer that their reality actually figures essentially within any temporal realism worthy of the name. This ground is not subjective experience or intuition, but rather the *ontological* ground, tracing ultimately to the ontological difference, that is indicated in the analysis of the contradictory givenness of time carried out above. As we have seen, the analysis points to a reality of time that is, on the one hand, contradictory, but on the other, (self-)given as the form of all temporal change and becoming. With their ontological and metalogical structure thus clarified and as part of a broader attitude of metaformal realism (see chapter 5), temporal change and becoming can be treated as fully *real*, while nevertheless avowedly paradoxical and *contradictory* in the way McTaggart's argument suggests.[67]

To defend the reality of change and becoming on these grounds is thus not to point to their source in the structure or activity of a constitutive subject; and neither is it to point, in response to B-series theorists, to further facts about subjects or experiences within the world that would purportedly be left out by the form of objective description of it that these theorists envision. It is, rather, to point out the structure of the phenomenon itself, as it gives itself, and in this very givenness, in essential relationship to the givenness of the world as such. As we have seen repeatedly in the analyses carried out in part 1 of this book, it is an essential part of the ontological interpretation of such phenomena as those of truth and time that these phenomena be formally-indicatively clarified in the specific structure of their ontological givenness or presentation. Such an indication is a matter neither of describing contingent facts, nor of giving an external definition of a phenomenon that is in itself and initially quite unfamiliar to us. As an instance of the hermeneutics of facticity, it is rather a question of the formal-indicative clarification of the phenomenon on the basis of the pretheoretic understanding of the phenomenon that we already have, itself grounded in our initially inexplicit understanding of being itself. Like sense and truth, time, change, and becoming are phenomena with which we are already in this sense "familiar," but which can be made more fully explicit through ontological interpretation. The formally-indicative analysis of them points out the way these structures are given in their own ontological sense, making explicit what is, at best, implicit in the way we ordinarily and initially experience them.[68]

Moreover, it is clear that the twofold paradoxical structure that results itself suffices to underwrite *temporal realism* in each of the three senses mentioned at the end of the last chapter. First, it allows for realism *about past and future events* arbitrarily far in either direction. As is clarified, indeed, by the first (cosmological) paradox, the possibility and sense of these events is not here seen as dependent in any way on the presence of the consciousness, subjectivity, or language that allows for them to be experienced, understood, communicated, expressed, or known; rather, it is grounded in the underlying ontological structure itself.[69] There is thus no inference to be made here to any form of antirealism about them.[70] Second, it allows for realism about the *presence* of the present: for instance, it allows a realist attitude about the content of utterances such as "It is five-thirty *now*" and further allows, as we have seen, that what is expressed by such utterances when true is, while not to be reduced to wholly non-indexical terms, not in any way dependent on an individual (or collective) consciousness or subjectivity. Third, and crucially in relation to the metaphysical tradition whose overcoming it allows, it underwrites realism about *change* and *becoming*. For, as we have seen especially in connection with the second (kairological) paradox, it allows for the contradictory linguistic determinations which signal genuine change to be accommodated, and even provided for, as direct *consequences* of the underlying and unitary formally indicated ontological-temporal situation.[71]

What, then, should be said to the B-series theory and its reductive account of the linguistic functioning of temporal indexicals (or token-reflexives), in terms of which it understands the *appearance* of change and becoming? First, without gainsaying the project of an analysis of the linguistic functioning of indexical terms, what is important to note here is that what the ontological investigation has in view under the heading of the analysis of the "now" is never *simply* the functioning of a linguistic term. Rather, much more broadly, it considers the concrete structural *phenomenon* within which this functioning takes place. What is in question in this analysis, for instance, is not primarily what a term such as "now" picks out in ordinary discourse, but rather the whole context of sense and significance in which it is possible for such an utterance of "now" meaningfully to take place at all. This analysis extends to the hermeneutic clarification of the ontological preconditions for there to *be* a kind of being—such as ourselves—capable of *saying* "now" in that it "has" something like a temporal present at all. Thus, just as Heidegger's own analytic of Dasein in *Being and Time* can largely be read as an ontological analysis of Dasein's essential and definitive "here," one can see the present analysis as carrying out, and carrying further, the implications of such an investigation for the temporal "now." The yield of the analysis in both cases is not any new addition to the inventory of (ontic) beings, or to the catalogue of facts about things in the world. It is rather the clarification of the being of the present on the basis of an indication of the ontic-ontological structure of its givenness.[72]

It is just here, as well, that an account of *truth* of the twofold kind sketched in chapter 2, above, proves helpful in underlying a realist doctrine of given time. As we saw, such an account looks simultaneously both toward *logical* and *structural* preconditions for the truth-value of sentences and toward *ontological* conditions for the unconcealment of entities. Both kinds of preconditions are irreducibly holistic, and they are jointly grounded in a unitary hermeneutic situation which is itself to be understood in ontological terms, and ultimately in terms of ontic-ontological difference. Through its formal-indicative significance, such an analysis is well suited, as I argued above, to account in realist terms for the ontological basis for the sense of sentences as well as things, the formal/logical structure of propositional truths as well as the hermeneutic/ontological structure of the presentation of phenomena as such.

What, then, if the analysis is applied to the truth of temporally indexical *sentences*, *sentences* involving past and future events, and *sentences* reporting and describing change and becoming? Applying it in this way involves considering not only the formal-symbolic "logic" of these sentences in the narrow sense that could be captured by a formal calculus but also, and just as irreducibly, that of the phenomena underlying them, those of the presentation of the present and of temporal change themselves.[73] Within the tradition of logical analysis, the possibility of such a formal/ontological consideration of the phenomena has been routinely excluded or dismissed. One reason for this dismissal has been the assumption that phenomena characterized by inherently contradictory structures cannot be real. Along with this, the distinction between the "domain" of the applicability of the formal-logical analysis of propositions and that of the empirical presence of entities has seemed to underwrite the correlative temporal distinction between the "timeless" or "atemporal" world of logical contents and the intratemporal world of the sensible. This distinction is already prepared and substantially suggested, as we have seen, by the Visitor's account of logic and being in the *Sophist*, and it is replicated in Frege's own picture of the timeless "third" realm. With the application of the formal/ontological analysis to the truths of temporal indexicality and change, however, it is replaced with a unitary, though paradoxical, account that provides for the time of sentences as well as entities in a single, ontologically motivated way, thereby overcoming the traditional distinction of temporal modes, as well as the whole development of the metaphysics it involves.

4. Consequences of the Structure: Generic and Paradoxico-Critical Interpretations

With the structure of temporal paradox clearly in view, we can begin to consider its consequences for the constitution, development, and limits of the historical tradition that Heidegger calls that of metaphysics or onto-theo-logy.

To begin with, it is significant to note that both temporal paradoxes (3a and 3b), besides the *Kantian* provenance which I have discussed, also have a *Greek* provenance and development. Both are discussed in their essential features, for example, by Aristotle in his "treatise" on time, *Phys-*

ics book 4.[74] Before this, both types of paradoxes are discussed explicitly by Plato.

The first kind of paradox, investigated for example in the *Parmenides*, the *Timaeus*, and the *Sophist*, relates the inherence of the infinite to the topic of the unity of the cosmological All, whereby the very structure of its *logos* always ensures "at least one more" and thereby tends toward the ultimate destitution of the One-All in a logically/structurally implicit unlimited many.[75] The second kind of paradox, investigated in the *Cratylus*, the *Philebus*, the *Theaetetus*, and again the *Parmenides*, is that of the thinkability of becoming and change, and more generally of the possibility of any thought at all of what is subject to the condition of temporal flux.[76] Even before this, the first type of paradox is clearly decisive in Parmenides himself, in producing the picture he maintains of a bounded but timeless cosmological One; and the second is more or less explicitly stated by his student and associate, Zeno, for instance in the "paradox of the arrow." Both types of paradoxes, in introducing a basic structure of contradiction into the thought of the One as such, underpin the later Plato's two-pronged attack on the Eleatic monism that treats being as the cosmological One-All and time and change as illusory and impossible. The development of this critique and the positive demonstration of the phenomena underlying its possibility allow Plato to rehabilitate and develop certain suggestions of Pythagorean ontology by expounding on the underlying problematic of the structural givenness of *number* to which that ontology responds.[77] At the same time, both paradoxes are then subsequently foreclosed or avoided, in both the Visitor's picture of logical/psychological *koinonia* and Aristotle's conception of temporal change as dependent on an underlying *dunamis*, by pictures that affirm the universal scope of the law of noncontradiction and thereby obscure the underlying paradoxical form of given time.

In this way, the foreclosure of the temporal paradoxes already suggested by the Visitor's positive position in the *Sophist*, and furthered in Aristotle's metaphysics, provides the specific historical condition for the constitution of the *onto-theological orientation* (in the narrower sense). Within this orientation, completeness and consistency are jointly maintained by means of the structural assumption of a superior logic of ultimate types and the temporal structure of *dunamis* and *energeia* at the basis of the metaphysics of substance. The temporal paradoxes are subsequently foreclosed just as decisively, albeit in a different way, in Kant's transcendental idealist picture and the forms of *constructivism* that arise from it and share its metaformal structure. For constructivism, as we have seen (chapter 5), the realm of phenomena is thought as the closure of a consistently thinkable totality, itself understood as the finitude of what falls within the bounds of constituted sense. Here the joint maintenance of consistency and completeness with respect to temporal phenomena depends not only on a presumed structure of superior ontological categories relating to *substance*, but also on the structure of the constitutive or transcendental *subject* in its representational relation to (otherwise unknowable) things in themselves. As we have seen, the suppression of temporal paradox within transcendental idealism itself takes two forms, corresponding to the two forms of reflexive paradox explored above. With respect to the *cosmological* form of paradox—that of the given time of the world as a whole—the constructivist repression takes the form of the familiar Kantian "solution" to the antinomies: that of the *denial* of the existence of the totality in question. With respect to the *kairological* form of paradox—that of the moment in its becoming—it takes the form of the subjectivist doctrine of faculties and synthesis as constitutive of the "I think" of transcendental apperception, culminating in the aporeatic doctrine of the schematism.

As we have seen, however, both of these forms of solution—the Kantian, constructivist one as much as the Platonic-Aristotelian onto-theological one—ultimately have the effect of denying the reality of (paradoxical) temporal change and becoming. Whether this denial takes the form of the Kantian attribution of the ultimate basis of given time to the faculties of the subject or the Platonic/Aristotelian attribution of it to the superior structure of dynamic categories, the effect is similar: each denial renders obscure the very reality of given world-time itself. As we have seen, on the other hand, this reality can be brought out if the underlying paradoxical structure of given world-time is itself clarified metaformally.

In order to grasp the real underlying structure of paradox, it is therefore necessary to take up one or the other of the orientations that unfold its consequences: the generic or the paradoxico-critical. How, then, do each of the two "post-Cantorian" orientations treat the paradoxical form of time? First, it should be noted that while both structures of temporal paradox I have discussed have been developed with a view toward a paradoxico-critical doctrine of them, both might also certainly be accommodated within the *generic* orientation. Within this orientation, in particular, limit-paradox has the significance of fundamentally articulating the boundaries of a specific, constituted regime of sense, while at the same time pointing beyond it to further possibilities of its problematic rearticulation or transformation from a perspective that is itself both (paradoxically) internal and external to the constituted situation in question. Taking into view the *cosmological* paradox of given time, the generic orientation construes it as demonstrating the sense in which each constituted "world," constituted by means of a determinate structure or regime of sense, bears within itself the possibilities for an ontologically based transformation of that regime. Such a transformation may unpredictably and radically transform the very possibilities for the appearance and phenomenalization of entities within such a particular world.[78] Here, then, worlds of appearance are essentially plural, whereas their possibilities of radical transformation are also seen as dependent on the superior global "ontological" perspective from which being delivers itself into finite phenomenality.[79] And similarly, taking into view the *kairological* paradox of temporal becoming, the generic orientation figures the privileged *site* of this possible transformation of sense as the punctual temporal one of the sudden and unpredictable "eruption" of the new, or (what it terms) the "event." Here, since it is ultimately such a punctual form that is responsible for all constitution of the sense and meaning of things, temporal change and becoming themselves may be seen as having the form of the periodization implied by the nonsynchrony of events, or their correspondence, across empirical time, in the superior medium of their infinite truth. The time of the world is thus understood as that of the circulation of the ultimately ontological and fundamentally *atemporal* form of the event and "its" truth—in the singular—through the specific worlds of phenomenality—in the plural—that it (in each case) locally determines.[80]

This generic interpretation of the ontological significance of temporal paradox has undeniable relevance for the ontological problematic generally. In particular, it characterizes well the structure of the formation and transformation of the successive *epochs* of the history of being within the history of metaphysics as onto-theo-logy. Within this history, as we have seen (chapter 4), the interpretation of beings is determined, successively, by one or another hegemonic "stamping" or phantasm, an (ultimately) paradoxical ontic-ontological figure that determines the intelligibility of entities as such for a particular historical world. Here, then, epochs are irreducibly in the plural, and the ultimate ontological provenance of the particular stampings is also at the root of their possible discontinuous transformation, what Heidegger himself specifies as (an aspect) of *Ereignis* or "the event." *Within* the ambit of this history, then, the constitution of specific worlds has exactly the temporal significance that the generic orientation would suggest: that of the periodic institution and transformation of determinate regions of total sense and phenomenality on the superior basis of an ultimately ontological dispensation.[81]

Nevertheless, as we have seen since chapter 1, Heidegger's "history of being" ultimately has in view not only the way that historical worlds are determined, within the history of metaphysics, by one or another interpretation of being in terms of beings, but *also*, and just as significantly, the form of the more general conditions of this *whole history*, as well as of its contemporary (in)closure. And here it is no longer simply a question of the ontological/temporal determination of worlds or epochs in the *plural*, but rather of the *underlying*, although hitherto dissimulated or foreclosed, form of the *time of the world*, in the singular and as such. By contrast with the generic orientation, which always views particular constituted worlds from a perspective of their essential incompleteness and in view of a superior ontological structure, the paradoxico-critical orientation alone is able to formulate and track the consequences of the underlying (inconsistent) temporal-paradoxical form from a purely *immanent* perspective. Here, for instance, there is no essential appeal to overarching "truths" indiscernible as such within a particular epochal situation, nor to a transcendentalist conception of the constitution of situations or worlds. All that is attempted is a clarification of the immanent phenomenological logic of the structural givenness of time, as it indeed already gives itself in the form of the present. It is thus to it that we can look for a clear formal and ontological unfolding of the implications of this original structure, and its essential relation to the determination of the temporal being of entities as such.[82] It thereby offers the resources (where the generic orientation by itself does not) to illuminate also the basic ontological and temporal determinants of the whole metaphysical regime of onto-theo-logy. Further, and for this reason, it also appears capable of illuminating the formal determinants of a transformed ontological/temporal situation to come, beyond or outside the contemporary (in)closure of this regime.

It is from this perspective that we can finally assess in broadest terms the ontological/temporal meaning of the law of noncontradiction itself. In particular, if the "metaphysical" assumption of the joint completeness and consistency of the ontic domain always rests on the suppression or obscuration of the more fundamental structure of temporal paradox, then the consistent ontic totality in general may be seen as the domain in which a "fundamental" prohibition of contradiction (or the assertion of its impossibility) is maintained in force. This prohibition itself produces the specific applicability and force of (noncontradictory) logic in relation to a totality of thinkable entities or contents. It is on this basis that the ultimate correspondence of the "logical" possibilities of thought and the "ontological" or metaphysical possibilities of existence is ensured in advance. But within the regime thus constituted, the prohibition itself will always then be thought in a double and ambiguous way: both as the "logical" impossibility of the contradictoriness of thought—the lack or absence of content of the "contradictory" thought or claim—and as the "metaphysical" impossibility of the contradictoriness of things, states of affairs, or the world in itself. This is the ultimate source, as well, of the ambiguities noted in chapter 4, whereby the law of noncontradiction always operates as the overdetermined debarring of what is anyway considered impossible for thought or reality. As a consequence and outcome of this, it is then also always possible to think the *self-identity* of the entities thought as subject to the force of the law of noncontradiction in a twofold way: both as *identical* "in reality" and thereby incapable of contradicting "themselves" *and* as *identified or identifiable* "in thought" and thereby incapable of *being thought* under contradictory determinations.

But this duality corresponds directly to the two formulations of the law of noncontradiction that Kant and Heidegger discuss, namely as involving temporal determination and as free of it.[83] If the law is thought as binding directly on the world of objects, as it is (e.g.) for Aristotle, it must involve the temporal determination, for its force then is to hold the unity of a (temporal) object in place by prohibiting the excess of self-difference in its real determinations that would necessarily double it. If it is understood, instead, as the highest or most general principle of the unity of judgment in *thought*, as it is for Kant, the temporal determination is not required, for its force is then only to render coherent the thought of the object, which is here understood as timeless in itself. From the present perspective, though, neither formulation is more "correct" than the other. Under the condition of the metaphysical interpretation of being as presence, what is essential to the maintenance of a structural correspondence of language to reality and ultimately of thinking to being is rather just the ambiguity between them. It is this temporal ambiguity which, as we have seen, actually underlies the onto-theo-logically assumed correspondence between the "logical" realm of thinkable contents and the ontic in general, and thereby produces the distinction between the temporal and the atemporal in terms of which they are metaphysically thought.

This is the correspondence that is, as we have seen (chapters 1 and 4 above), already invoked by the Visitor's picture of the psycho-logical *koinon* of the great types, both as the thinkable and the real determinations of being in itself. It is further consolidated and maintained, on Heidegger's reading of the history of metaphysics, by the standing assumption of an identity whose ultimate form is the *idea* as the common or general, or again by the Kantian structure of the constitution of the categories of phenomena by means of the subject's synthetic activity and its transcendental apperceptive unity. Finally, it is the idea of an active correspondence between the identity of things and their identification in thought—the correspondence which is ultimately understood, through all of these configurations, as the self-correspondence of the rule capable of repeating itself endlessly as the same. It is this thought of self-correspondence as infinite thinkable repetition which figures in one way, as we have seen, in Kant's conception of the schema of substance as the ultimate form of persistence, and in another in the conception of the rule as a "rail laid to infinity" which Wittgenstein challenges. But it is fatally problematized by means of the ontological-grammatical exhibition of the more original structure at the temporal basis of presence. This exhibition evinces the ultimate emptiness of the constitution of identity, and hence the consistent and total correspondence between thought and being, that the "metaphysical" or onto-theo-logical idea of presence tries to found. Behind this metaphysically assumed correspondence, the original structure of temporal/reflexive paradox is thus shown as both the underlying ontological basis of this idea of presence and the condition under which it can no longer be maintained in force.

7

Gestell, *Ereignis*, and Eternal Return

I have argued that the illumination of the temporal paradoxes provides a figure of the "beyond" of the closure of metaphysics in the sense of onto-theo-logy, and hence a condition for its possible overcoming which is formally indicated in the metalogical results. But it is the contemporary actuality and ubiquity of *information technology*, itself the direct technological yield of early twentieth-century logical/symbolic formalism, which points most directly to the conditions of such an overcoming in contemporary life. As I argue in this chapter, with the radical and pervasive development of information and computational technology characteristic of our time, the history begun with the Eleatic Visitor's identification of the *dunamis* of the logical *koinonia* as the psychological/logical structure ultimately responsible for the correspondence of thought and being reaches an inherent completion. For with this development, the history of the logical *dunamis* is completed in the complete *realization* of "logical" structure as the ubiquity of the total claim of *computation* and *computability* and the total shaping of practices, economies, and ways of life in its form. Here the logical *dunamis* is thus, finally, no longer grasped as the principle of an individual-psychological mediation, in the *psuche* of the individual subject of judgment or knowledge, between beings in themselves and the forms in which they are given to be thought. Instead, with the totalization of "instrumental" and computational calculability, this *dunamis* is immediately and fully *actualized* as the "totalizing" reality of effectively organized "social" practices, institutions, and forms of life. In this actuality of "logical form" as realized technology and technological practice, *techne* as the formerly neutral or abstract "third term" between being and thinking thus becomes directly *realized* and actually and generally "effective" as the unified form of all comprehensible communication and action. With this, as well, what was previously known as the notional or abstract "force" of reason is instrumentalized, formalized, and brought *into* effect as the total and unified reality in which all beings and their thought are simultaneously lodged. This is why, as Heidegger himself often suggested, the "culmination" of metaphysics takes the form of the ubiquitous regime of *completely* effective actuality or "reality [*Wirklichkeit*]" which no longer allows any reserve of *dunamis* within which thoughtful and critical reflection can ultimately shelter, but only the full and direct accomplishment of effective force.

This drive to the completion of effectivity is visible in the typical character of *totality* that structures the most distinctive and ubiquitous problems of the contemporary "global" situation, whether these are further determined in the registers of the political, the "economic," the sociological, the ethical, or the ecological. They are problems about the constitution of the *whole*, negotiations of the boundaries of what is seen as *all-inclusive*, and border questions at the limits of the logic of *total systems* that claim and enforce global comprehensiveness. This extends, as I shall briefly argue here, to the inherently "totalizing" logic of contemporary global capitalism itself, which constantly drives toward completeness in its structuration of contemporary forms of life.[1] These problems are also, just as much, characterized by the presumed effectiveness and force of what is (ambiguously enough) called "technology," which I here understand in a sense that preserves the link of the term with the Greek *techne* as well as with the later-developed sense of "technique" as a *regular* procedure carried out by means of essentially specifiable rules. In the light of these problems and in view of the situational factors that determine them, the original question of the relationship of thought and being also takes on a distinctive shape. This shape is conditioned as much by the radical discoveries of metaformal insight that make the structure of the infinite available to mathematical thought, in our time, as by the specific technologies and material practices of computation and "information" that follow directly from these discoveries.

Heidegger's own conception of a contemporary "culmination" of metaphysics, marked in the technological "exhaustion" of the possibilities of thinking first instituted at the Greek beginning, is undoubtedly indispensable here. Nevertheless, the deepening of the ontological problematic on metaformal grounds that is also here suggested already suffices to articulate the significance of ontological questioning for the problems of the contemporary, and of what is "beyond" it, or "to come," in a different direction than any indicated by Heidegger himself. In fact, the suggestion of a broader metaformal conditioning of the metaphysical tradition is already essentially made as soon as the structure of the formal problematics of time and sense that we have pursued are indicated. These problematics themselves are configurations to which the thinking of "being as presence" must itself be originally responsive, and to which this thinking again opens up at the determined point of the closure or exhaustion of its determining rubrics. From this perspective it is no accident that, as I have argued, they also indicate original formal-ontological problems that plausibly underlie the structure of "natural" time just as much as they do "historical" time, articulating the formal and metaformal conditions ontologically characteristic of given time generally and as such. Here, accordingly, what is at issue with respect to the determined "history of metaphysics" is no longer the task of a deconstructive "retrieval" of what is thought at its origin and then progressively concealed within it. Nor is the task one of safeguarding, during or after the period of the totality or exhaustion of this history, what within it is, as ultimately determining for it, anyway purported to be its greatest exigency (namely "being itself"). It is, rather, just a question of understanding how this (relatively) determined history can be seen to communicate *formally*, at a certain point, with its *own* broader determining conditions in the underlying problems to which it points, and thereby how, with its determined closure, it also and in the same moment points to a broader "outside."[2]

1. The Essence of Technology and the (In)Closure of Metaphysics

In a number of texts of the 1930s and 1940s, Heidegger describes the characteristic and dominant configuration of contemporary thought and practice as a "machination [*Machenschaft*]" which (circularly) aims to order all beings with a view only to their ever-greater orderability and thereby tends toward the assumption of their universal *representability* and *calculability*.[3] Heidegger understands the contemporary dominance of *Machenschaft* as itself the expression of the necessarily *final* epochal determination of the being of entities within the series of configurations that make up the history of metaphysics as a whole. With this dominance, the "possibilities" of metaphysics are in a certain way completed or exhausted. This exhaustion is marked in the *completion* of the progressive withdrawal of being itself from thinkability that has characterized the development of metaphysics since its inception. In previous epochs, entities remained stabilized by means of particular epochal referents which, though elevating a particular kind of being to the rank of a universal standard, nevertheless still preserved in these standards something of the thought and intelligibility of being itself. By contrast with this, however, in the contemporary epoch of the universal calculability and manipulability of entities without any such stabilizing referent, this thought and intelligibility has completely disappeared. This provides the condition under which metaphysics, in the sense of onto-theo-logy, is definitively and irreversibly exhausted. But it may also, according to Heidegger, provide the positive condition for the transformation into a different condition, one in which being can once again be thought and experienced as *Ereignis.*

After the 1940s, Heidegger typically replaces the discussion of "machination" with that of "enframing [*Gestell*]," which he describes (for example, in the well-known 1953 lecture "The Question concerning Technology") as the "essence" of technology. Such an essence is, he says there, not itself any technological thing, nor something determined or determinable as an outcome simply of human activity.[4] However, at various places in his late work, Heidegger gestures toward an enigmatic relationship, indicated as to be "looked into" or "looked forward to" in the further development of ontological thought or the thought that comes after it, between *Gestell* and *Ereignis* itself. One formulation of this is given in the Le Thor seminar of 1969:

> An excellent way of approaching *Ereignis* would be to look into the essence of enframing [*bis in das Wesen des Ge-stells zu blicken*] insofar as it is a passage from metaphysics to another thinking . . . for enframing is essentially ambiguous . . . Enframing is, as it were, the photographic negative of *Ereignis.*[5]

With this formulation and others like it, Heidegger seems also to indicate a curious relationship of the *totality* of the history of metaphysics—thought as its "culmination," or the "exhaustion" of its possibilities in the contemporary dominance of *Gestell*—with something outside or beyond it that also determines it. This is *Ereignis,* which grants or ontologically conditions each of the determinate configurations of metaphysics and the whole "epoch" of presence itself. The relationship suggested here between the final of these configurations and this opening is neither extant nor assured; it is, rather, to be glimpsed, anticipated, or prepared, recommended as a *possible* path for thought to come, its own provenance dependent ultimately on the granting that *Ereignis may* allow. Nevertheless, in that *Ereignis* is thought, as the very event of the happening of an "other" to the history of metaphysics, what is, here, at least indicated, appears to be the actual form of a *thinkable* relationship between all that is enclosed in this history and what insists beyond it as its condition and exterior.

The problem of this relationship then takes determinate form as the problem of the "present" time of technology in relation to the time that, surrounding and determining it, also can be indicated on the basis of a formal-ontological thinking of it. This thinking would itself necessarily be, in one way, the repetition of a thinking of time or of the possibility of a future that is "determined" by the essence of technology—and *in particular* is determined through and at the limits of calculation and calculability and the configuration in which it relationally places thinking and being. And indeed it would not go too far to say that the forms in which a "future" can today be envisaged, the forms in which our age knows itself in relation to what it thinks as its possible "beyond," are so many figures of this very relationship between a technologically determined thinking of time and the "event." Today, this relationship is envisioned in alternate forms of anxiety and celebration, for instance as the technologically inevitable moment of global catastrophe or as the equally determined "singularity" in which technological thinking finally overcomes itself and transforms itself, no longer needing the assistance of the organic "human" powers of thought, into something unforeseeable by means of them.[6]

As it develops after the 1940s, Heidegger's thought of the preparation for an overcoming of technological enframing is (familiarly) marked by a waning of the rhetoric or figure of decision, and becomes largely characterized as the thought of "*Gelassenheit*," or of the peculiar "letting" beings be which is to allow once more the possibility of their *be*ing in an originally disclosive way. This is the thinking, apparently at the most extreme possible distance from technology, enframing, and its characterization of beings in terms of acting and effectivity, of the "saving power" of the original unity of *techne* and *poiesis*, and of the "fourfold" in which the counterplay of earth and sky, man and gods, holds open the place of disclosure and granting as such. The suggestion of this attitude as the successor to a modern "enframing" attitude toward beings has been both endorsed on Heidegger's behalf and, elsewhere, criticized as fundamentally regressive and nostalgic.

But without disputing any of these terms in which Heidegger ventures to think or propose the possibility of what cannot yet, *sensu stricto*, even be thought (since its thinking would *already* accomplish it), it is possible and coherent to ask whether there are more specific terms to be found, within the concrete factical interpretation of the determined "present," for the indication or demonstration of what *surrounds* the history of metaphysics as its prior condition and more general exterior. How, in particular, in the thought and reality of the "consummation of metaphysics," is a certain "outside" already indicated or *readable* in today's determinate structural, temporal, or ontic-ontological forms? The question asks (with Heidegger) about what *in onto-theo-logy* points to whatever is "beyond" its own totality, but *also* about what points (formally or ontologically) to this "beyond" *in the realized idea of totality itself*, as this idea gives itself in the constitutive forms in which it also shapes and regulates contemporary thought and practice.

In preliminary response to this question, it is relevant to consider in a specific way the ever-more-pervasive technologies of information production, distribution, and exchange which today encircle the globe and affect practically every life on the planet, through the possibilities of communication, manipulation, and economic capture they afford. It appears possible, in particular, that this dominance witnesses a specificity of information technology itself—as opposed to the older forms of mechanical or industrial technology that Heidegger himself most often refers to—in articulating the form in which the metaphysical tradition today reaches its most definitive culmination and exhaustion.

Heidegger himself sometimes anticipates the contemporary development and predominance of information technology, even in significant detail. For instance, in the 1964 essay "The End of Philosophy and the Task of Thinking," in connection with the "role of logic as symbolic logic and semantics" he points to "cybernetics" as the "new fundamental science" that will soon "[determine] and [regulate]" all other sciences.[7] It is equally significant in relation to this to note the central connection Heidegger draws, already in the 1930s, between the dominance of machination and the character that beings attain within it of being universally countable or *calculable* [*Zahlbar*]. Thus, for instance, in the *Beiträge,* in close connection with the contemporary dominance of a reign of "calculative thinking," "machination" is specified, more basically, as the interpretation of all beings as "representable and represented" and as such "on the one hand, accessible in opinion and calculation [*in Meinen und Rechnen*], and, on the other hand, providable in production and implementation."[8] In this way of thinking, all that counts "as being [*als "seiend"*]" is "what is or can be the object of a lived experience [*das Er-lebte und Er-lebbare*]" in being able to be brought before one in representation.[9] To this conception of being as what is representable corresponds the apparently *unlimited* representability of beings in machination, which is itself guaranteed by their unlimited measurability and calculability. Here there is no limit to what can be given *for* representation, since there is "*nothing* that is, or could be, given" *to* calculative representation "*as a limit.*"[10] This unlimited calculability means that "the incalculable is merely that which has not yet been mastered in calculation [*das in der Berechnung noch nicht Bewältige*] but is in itself also to be captured some day [*an sich aber einst auch Einzufangende*]"; there is, accordingly, "in no way . . . anything *outside all calculation* [*also keineswegs das Außerhalb jeder Rechnung*]."[11] And because of the certainty about beings which this in-principle calculability of everything provides, "the question of the essence of truth" is itself "no longer needed."[12]

What comes into view here in the specific character of contemporary technology as total, calculative enframing is indicative of, because fundamentally rooted in, what is for Heidegger an (perhaps *the*) *original* and determinative structure of metaphysics itself: namely, the structure of logic as a regular *technique*. Here the measurability and calculable manipulation of all beings is itself assured on the ultimate basis of the applicability of regular methods of their representation and computation, the very methods that are subsequently captured and universally promulgated in the actualized form of the digital computer. The "contemporary" situation of global information realizes this regime all the more fully, insofar as it allows and propounds the dominance of calculable technological and economic manipulation, and the assumption of universal effectiveness that underlies and supports it, around the world and with respect to all its (thereby determined) forms of life.

It is relevant to this that the unitary structural basis of all contemporary forms of information technology in the basic formal architecture of the digital computer also, and by its own inherent structure, in fact demonstrates a formally necessary ontologically significant phenomenon of *undecidability* and *ineffectivity*. The connection is in view in Alan Turing's remarkable 1936 paper, "On Computable Numbers, with an Application to the *Entscheidungsproblem*."[13] In the paper, Turing simultaneously created the logical structure of the modern electronic computer by defining the architecture of a machine capable of carrying out any well-defined algorithmic procedure *and*, in the very same argument, demonstrated the existence of (countlessly many) mathematical problems that essentially *cannot* be decided by any such procedure.[14] Along with this, and as another consequence, Turing's result demonstrates the existence of infinitely (uncountably) *many* "uncomputable" real numbers: numbers whose decimal expansion cannot be determined by the reiterated application of *any* finitely statable rule.[15] A further corollary (derivable by partial means of an auxiliary result)[16] is a form of Gödel's first theorem itself: for any system of sufficient complexity to capture arithmetic, there will be an arithmetic sentence that it cannot prove or refute, provided the system itself is consistent.

Here, then, the very logical architecture which, as realized, provides the actual concrete basis for the whole contemporary regime of "universal" calculability and decidability is also shown to entail the inherently necessary existence of problems that are undecidable and quantities that cannot effectively be computed, in a strict and rigorous sense. With this demonstration and in it, the very idea of total effectivity, as that of the *universal* procedural determinability of all beings by regular computational means, is negated in its general form and shown ultimately incapable of founding the assumed "total" regime that it is invoked to support. As is shown by Turing's own results and verified by the other metaformal structures and results we have considered, this means that any demarcation of a domain in which procedures can be considered effective and always capable of producing determinate solutions with respect to well-posed problems must always be understood as taking place only within a *broader* realm of *ineffectivity*, wherein the smooth functioning of these processes cannot be guaranteed and the procedural availability of solutions is not and cannot be generally assured. This may be seen as extending as well (and in an eminent sense) to the *total* demarcation of the "contemporary" regime of universal calculability and decidability, which Heidegger treats as the contemporary dominance of machination. As such, this regime appears to bear in the underlying structure of its material/technological basis the formal conditions for an overcoming of the ontological assumption that maintains it.

If Turing's result thus shows in formal detail that any determination of regular effectivity—and also all that assures and guarantees it—communicates with a broader undecidability which is its general surrounding, then it may be possible to draw from this some critical considerations relevant to the characteristic and general forms and practices of contemporary technological, social, and political life. Specifically, it is here apparently possible to envision a formally based critique of these forms and practices that operates by problematizing and challenging the *claim* or *assumption* of general effectivity that underlies their application and force. Here the formal indication that appears to be derivable from the actual metaformal results rejoins the Heideggerian ontological investigation at the point where that investigation seeks to provide terms to challenge and question the contemporary dominance of "machination" and its characteristic assumption of the total manipulability of entities. For as Heidegger points out, the basis of this assumption is just the claim of their unlimited possibility of being represented, itself grounded in the totalization of their calculability. If, then, there is a formally demonstrable incalculability inherent in the very thought of regularly calculative procedures itself, then the general suggestion, already made in Heidegger's text, of a specifically determined "limit" and "outside" to the contemporary technological regime here receives significant confirmation and is put on a formally rigorous basis.

What are the implications of this for the "ontological" problematic of sense, truth, and time and for the critical problems of thought and action that face us today? As we have seen over the chapters of this book, the metaformal dynamics of these phenomena as they are indicated in their givenness lead to the demonstration of an inherent *undecidability* of sense (chapter 4) which is closely connected, in the context of the ontological determination of beings as a whole and as such, with an ultimately *paradoxical* structure of given time (chapter 6). Both results can themselves be connected to the procedural undecidability that Turing demonstrates: the first, in that it bears witness to the consequences of computational undecidability and ineffectivity for the structure and temporal character of languages in their semantic relation to the world; and the second, in that the paradoxicality of given time as the form of presence renders it impossible for temporally determined sense to be *consistently* decidable. Synthesizing the two results and drawing out their critical implications, the formal analysis of the consequences of Turing's results thus appears to bear witness to an inherent *undecidability of the sense of temporal becoming*, which appears henceforth to render impossible any univocal *determination* of the meaning of technological or social change, "progress," "development," or transformation. If, in particular, it is impossible (given the metaformal results) to *decide* the sense of technological or social "development" (whether this is thought in gradualist terms, with a preestablished goal or telos already more or less in view, or indeed in terms of "radical" transformation, evental breaks, and the like), then it is also impossible to found a post-metaphysical politics on the hope for a redeemed future or an improved state of the world to come. What is necessary, instead, is to draw out and sharpen the implications of the recognition that the sense of the future is (fundamentally and radically) undecidable, in the sense that it cannot be determined, secured, assured, or achieved by means of any regular or instrumental procedure.

In the context of contemporary critical thought about global capitalism and the total regime of thought and practice that it structures, this has specific consequences, some of them at significant variance from those that are usually accorded to Heidegger, or to a "Heideggerian" thinking of politics or technology critique. There is no suggestion here, for instance, of the usefulness or even possibility of a marginal withdrawal from the prevailing technological forms of organization and effective structure. Nor is there any need here for, or benefit to be derived from, an essentially nostalgic politics of *return*, whether to an earlier premodern moment thought as innocent of the distorting effects of technological instrumentalization and production or, indeed, to one before the "beginning" of the "history of being" itself.

More positively, however, what is suggested on the level of critique in relation to contemporary capitalism is the structurally necessary failure of its claim to noncontradictory totality: its claim to capture and accommodate, without contradiction or fundamental antagonism, the totality of planetary forms of life. If the undecidability of temporal sense suggested here can indeed be generalized and applied to the structure of contemporary life and practice, it shows, in other words, that the characteristic attempt of capitalism recurrently to produce and then capture its own specific "outside" by means of regular instrumental procedures must necessarily fail. For the general ideological claim at the basis of this attempt, that of the *universal* noncontradictory calculable decidability of beings, itself fails as a consequence of the formal results. This suffices to show, in formally motivated terms, the futility of the claim of global capitalism to totalization, and hence demonstrates the emptiness of the ideological terms and practices that (whether by means of ever-increasing economization, instrumentalization, or "securitization") today seek to maintain and extend this claim.[17]

2. Temporal-Paradoxical Overcoming of Onto-theo-logy

Over the last chapters, we have seen formal grounds for beginning to think an original structure of paradox that is formally indicated, both as the infinite given time of the cosmos and as the paradoxical form of the instant. As I have argued (chapter 6), this paradoxical infinite structure of the instant and of the world is itself indicated as the *real underlying structure* constitutive of given time, as soon as we take seriously the metaformal realism that results from reflection (chapter 5) on the availability of the actual infinite to thought. If Heidegger does not follow out these connections in detail, as we have seen, he nevertheless repeatedly indicates a thinking of the original form of time that would be at or "beyond" the formally indicated specific limits of calculation as such. This is the original form of time that, on the Heideggerian analysis of it, also surrounds and conditions the specific history of metaphysics. The metaformal analysis of it, followed out (as we have attempted to do here) in terms of the original paradoxes structurally constitutive of time as it gives itself both cosmologically and kairologically, thus structurally points to this more "general" time *within* which the history of metaphysics, stretched between its Greek inception and its contemporary closure, "takes place." For this reason, the structure of an originally undecidable given time can be seen as *the specific limit* and also the *structural overcoming* of the onto-theo-logy which recurrently organizes, within the history of metaphysics, the interpretation of the being of beings.

As we have seen since chapter 1, onto-theo-logy grounds both "from outside" and "from inside" or "from above" and "from below": both by selecting a superior ontic referent that sets a standard for the intelligibility of beings *as a whole* and by selecting a correspondent ontic principle for the constitution of beings *as such.*[18] Both aspects of grounding together determine, in each case, the form in which beings are understood and experienced, and both have a more basic original temporal determination in the form of the present. If, on the other hand, the original structure of given time is thought as paradoxical in the twofold way I have suggested, both with respect to the kairological paradoxical structure of the moment and with respect to the cosmological paradoxical structure of the totality, then *both aspects of onto-theo-logical grounding are overcome in favor of their more original infinite-temporal paradoxical conditions.* It suffices for this that onto-theo-logy's characteristic attempt to define beings "from above" by means of an elevated hegemonic referent is *replaced* with the cosmological temporal paradox (3a above), and its characteristic attempt to ground beings "from below" by means of a description of the conditions of their presence is *replaced* with the kairological temporal paradox (3b above). In each case, it is the paradoxical temporal structure which is itself now to be seen as grounding the being and sense of things. It does so, in each case, by conditioning the form of the presencing of beings through the way it conditions the temporal present itself. This yields a transformed twofold conditioning of the being and sense of beings, one that, like onto-theo-logy, continues to ground "from above" and "from below," but does so now by essential reference to the paradoxical conditions of temporal presencing *as such* rather than by means of any kind of theological or ontic grounding. With this, there will henceforth be no theological grounding of beings "from above," since the whole is not given from the perspective of the theological elevation of a supreme being as its consistent determining principle. It is now, rather, "given" as the infinite and paradoxical extent of cosmic time, unbounded (by a noncontradictory limit) in both directions. Similarly, it will no longer be possible to stabilize beings "from below" by means of the specification of necessary existents (such as Aristotle's *stoichea,* substances, monads, or constitutive beings-in-themselves of whatever type). For the temporal modality of the existence of these privileged beings—that of constant, standing presence—is itself now replaced by the kairological paradox of the moment, according to which what becomes is, as such, always becoming something else along with the basic form of given time.

This inscription of the temporal paradoxes at the level of the grounding of beings implies that "beings as such and as a whole" can no longer be stabilized in the form of a total and *consistent* interpretation. But this is just to say that "being" is, henceforth, itself no longer given within the closure of a (jointly complete and consistent) interpretation *of it* as presence. It is in this way that the (in)closure of onto-theo-logy and of the history of the interpretation of being as presence are decisively marked in the formal indication of the underlying paradoxical form of given time.

Henceforth, more generally, it will no longer be possible to stabilize beings or their becoming by means of ontic referents. In the condition of (what Reiner Schürmann calls) "ontological anarchy" (cf. chapter 4) that results, the destitution of all such referents is formally indicated in the infinite-undecidable structure of given time itself.[19] As we have seen, this general destitution of ontic referents in our time is also an inherent aspect of the contemporary dominance of technology or *Gestell*, in which beings become characterized by the general assumption of their calculability, without this assumption being underwritten any longer by any single referent or centering standard of presence. But in pointing to the broader conditions of onto-theo-logy itself, the formal indication of the broader paradoxical structure of given time also points to the more original conditions under which it was historically first *possible* to pursue a solution to the original problem of time and being by *logical* means. These are the conditions which, we have seen reason to think, presented themselves to Plato, and determined the form of the solution presented by the Visitor in the *Sophist* in terms of the *koinonia* of the great types, and again determined Kant's thinking about the basis for the categories and their application to temporal phenomena in the doctrine of the schematism. The solution, in the categorical form in which it was later further consolidated by Aristotle, or the transcendental form in which it was thought by Kant, persists and continues to determine the tradition's thinking about logic and presence until the moment at which the metalogical problematic itself reveals the necessary communication of logical categories and procedures with the broader infinite-undecidable temporal structure in which they inhere.

Given the metaformal development of the temporal paradoxes, time is then itself no longer determined in the medium of the noncontradictory presence of the present or as the thought of its consistent, self-similar repetition. This means that "being" is itself no longer given—nor can it be—within the form of a complete and consistent presence, but is temporally given only in, and as, the twofold paradoxical form that contests and undermines any thinking of it in these terms. With this, we have thus indicated here the broader temporal and metaformal situation underlying onto-theo-logy or the "metaphysics of presence" itself and as a whole. This situation appears both to condition this metaphysics as its underlying more general structure and *also* to become rigorously thinkable at a certain determinate moment of its internal development, namely the moment "we" are living "today."

3. Metaformal Figuration of Temporal Paradox: Eternal Return

The specific temporal overcoming of metaphysics is thus formally indicated through the development of the twofold temporal paradox of given time. This points the way to a condition or circumstance that is "post-metaphysical" in the sense of situating itself beyond and outside the onto-theo-logical metaphysics of presence. But how is this more radical temporal situation "to come" to be positively figured, thought out, and inhabited in a futural thought and practice, a specific "way" of being *in* time?

As we have seen (chapter 5), the history of onto-theo-logy, in both its "onto-theological" (in the narrower sense) and constructivist forms, is essentially conditioned by the idea of a joint consistency and completeness in the description of beings as such and as a whole. For this reason, to comprehend the temporal condition beyond its closure is to occupy one or both of the two post-Cantorian orientations: the generic one or the paradoxico-critical one. And as we have also seen (chapter 6), while the generic orientation indeed offers ways to think about the history of metaphysics itself from (something like a) "post-metaphysical" perspective, in order really to demonstrate and develop the implications of the underlying temporal paradoxes themselves, it is necessary to see them from the paradoxico-critical perspective that draws out the implications of the givenness of infinite time from a perspective wholly *within* the time thus given. The assumptions of this orientation which make possible the demonstration of the temporal paradoxes themselves are relatively straightforward: (i) that given time is *real* (and not ideal, illusory, subjective, etc.); (ii) that it is *universal* (i.e., there is nothing that is given as or from "outside" time, and thus, as a corollary, that time is given only from a *perspective* that is essentially itself "in time"); and (iii) that it is *actually* (and not just potentially) *infinite.*

As I argued in chapter 6, these three assumptions are sufficient to yield the two forms of temporal paradox, the cosmological one concerning world-time as a whole and the kairological one concerning the form of the "now." Given just the paradoxes, however, we might justifiably wonder what guidance they provide to how we can actually *live* in the temporal situation they appear to indicate. What would it be like to *inhabit* such a time of unlimited, paradoxical becoming, wherein nothing in the world is any longer assured to maintain its identity by means of a static form of time as the repetition of the self-same instant in presence, and the temporal sense of the world itself is radically undecidable?

At first glance it might seem that there is no *direct* way to figure such a temporal situation, since it characteristically involves, as we have seen, that it is in fact impossible temporally to figure either the world itself or its becoming both *wholly* and *consistently*. However, there is in fact a familiar and suggestive figure which formulates *one* way of thinking of the implications of all three assumptions stated in the paragraph before last. It is one which, moreover, captures the temporal situation that they point to in the specific form of an overall picture of the given time of the world, one with significant implications for thought, action, and practice. It is what Nietzsche formulates as the "doctrine" of the *eternal recurrence of the same*.

Near the beginning of a 1937 interpretation of Nietzsche's doctrine of eternal return as presented in the narrative of *Thus Spoke Zarathustra*, Heidegger suggests that it may be seen as turning on the specific presuppositions of the *infinite actuality of time* and the thinkable *finitude of things*:

> Nietzsche summarizes an essential thought concerning his doctrine so succinctly here, in the form of a question, that it is hardly comprehensible on its own, especially since the requisite presuppositions, although mentioned, do not really become visible. Those presuppositions are: first, the infinity of time in the directions of future and past; second, the actuality [*Wirklichkeit*] of time, which is not a "subjective" form of intuition [*die keine "subjektive" Form des Anschauens ist*]; third, the finitude of things and of their courses [*und dingliche Abläufe*]. On the basis of these presuppositions, everything that can in any way be must, as a being, already have been [*muß alles, was überhaupt sein kann, schon als Seiendes gewesen sein*]. For in an infinite time the course of a finite world [*der Lauf einer endlichen Welt*] is necessarily already completed.[20]

These are, however, just the three assumptions mentioned above, as the basic determinants of the paradoxico-critical orientation and its temporal significance of the demonstration of the kairological and cosmological paradoxes of given time. It is easy to see, moreover, how they receive independent motivation and confirmation within Nietzsche's system of thought. That time is *infinite* (assumption iii above) results (in the context of Nietzsche's system) from the assumption that the world has no beginning and no end—an assumption itself motivated by Nietzsche's atheism and the consequent requirement that there be no outside *boundary* to world-time as a whole. That *everything* is temporally conditioned (assumption ii) results from Nietzsche's interpretation of the basic character of being as *becoming*; it follows, as we have seen (chapter 6), that (since there is then no "transcendent" perspective outside time from which to contemplate it or observe it) any possible perspective on the given time of the world is itself temporal. And finally, that time is *real* (assumption i) follows from the Nietzschean critique of representational subjectivism, which Heidegger himself develops and strengthens (as we have seen).

Given these assumptions, it is also easy to see how something like the structure of eternal return can be motivated on the level of the *time of the world as a whole.* This time has no end and (*a fortiori*) no beginning. But everything that happens, happens *within* it, and is as such finite. That is, every event or process—*including the process of the total reflection of the course of the world as a whole in (the activity of) finite thought*—is (as such) finite and thus limited. But this—the totality of all finite processes in time—includes *also* any (thinkable) "process," or "course" of the world itself, and as such. For if the world is thinkable *as a whole,* it must also be able to be thought, as such, *in* finite time.[21] Thus the finitely determined course of the world *as such* (that which Heidegger's gloss on Nietzsche's argument presents as the "course of a finite world") must have *already* occurred—as a whole and up to the end of its process—*within* the infinite time that has already been "given" "up to" the present (and, indeed, not only once, but infinitely many times).[22] The argument turns essentially on the underlying paradox that the temporal totality of the world is both *infinite*—in the sense that every limit can always be outstripped—and *finite*—in the sense that it is *thinkable* as a whole, in finite time. As such it captures and figures in a basic way the paradoxical situation of "finite" and temporally situated thought itself in relation to the infinite givenness of thinkable time. But this is just the inclosure at the basis of the cosmological paradox of given time (as developed in chapter 6). *One* figuration of a temporality determined by this paradox is thus that of an eternal recurrence of the (thinkable) "course of the world" as such.[23]

It may be less evident, at least at first, that Nietzsche's doctrine is *also* very closely connected to the kairological paradox of the moment. But as Heidegger himself suggests, the connection can be brought out, this time, and in a way that also brings out the ontological significance of what Nietzsche calls *amor fati,* by considering the metaformal and temporal implications of the very *perspective* from which Nietzsche (or Zarathustra) speaks in announcing it. By formulating essentially the argument considered in the last paragraph, Heidegger suggests, Nietzsche is able to propose the doctrine of eternal recurrence as a fundamental position relating *beings as a whole* to their temporal condition. But he is only able to do so by first placing *himself* in the "moment" in a decisive way. In the placing, the thinker of the thought of the eternal recurrence is "transposed to beings as a whole [*in das Seiende im Ganzen*]," or placed under the particular condition of the way in which beings as a whole appear in the "gateway" of the moment in which future and past meet.[24]

In particular, in book 3 of *Thus Spoke Zarathustra,* the doctrine is presented by Zarathustra to a dwarf.[25] Zarathustra indicates a gateway on which is inscribed the name "moment [*Augenblick*]"; from the gateway stretch two eternal paths, ahead and behind. Whatever *can* happen has already happened on the path behind, and will happen again on the path running forward. All things are thus "bound together [*fest . . . verknotet*]" in such a way as to be drawn behind them by this gateway and must occur again, even the gateway itself must therefore occur once more. On Heidegger's interpretation of the passage, "time and eternity" are here viewed in an image oriented from the "moment" or "now."[26] The image is not, though, directly to be understood as Nietzsche's final theory, but rather just as the "riddle" that presents a question. The "doctrine," in particular, is first posed as a question to which the dwarf appears readily to have an answer: the dwarf asserts the circular form of "time itself [*die Zeit selber*]."[27] However, the question is repeated later on in the passage, this time as the question of whether *this* moment—the moment of Zarathustra and the dwarf and their questioning itself—must not be repeated eternally. And this time the dwarf has no answer: with this second formulation, Zarathustra himself becomes "afraid" of his own thoughts, and he has his vision of a young shepherd being bitten in the mouth by a snake, the head of which he must bite.[28]

The difference between the two formulations, according to Heidegger, lies in the fact that the second question is posed, not as a question about the general form of time, but on the basis of the present *moment,* as the question of the recurrence of *this* moment itself. With this particular kind of questioning, one is required to take a "stance of [one's] own within the 'Moment' ['*Augenblick*'] itself, that is, *in* time and its temporality."[29] With this figure of the "moment" and its structure, the moment is itself thought as the form of unlimited becoming, or as the "gateway" through which all things with fixed and stable identities tend to become something else. As we saw in chapter 6, though, to think the moment this way is just to think it as the *given form* of the reality of temporal becoming as such. Furthermore, to think it in this way already implies (in conjunction with the cosmological and ontological assumptions given above) that it is thought in terms of its paradoxical self-givenness. This thought already requires the (second) form of temporal paradox, the kairological paradox of the moment, whereby the reflexive structure of time's own self-givenness is grounded in the ontological structure of unconcealment—or truth—as such.

Furthermore, the nature of this connection resolves what otherwise presents itself as a puzzling aspect of Nietzsche's presentation of the eternal return: namely, that although it is on the one hand certainly intended as a *cosmological* doctrine that is seen as true, it is also, just as importantly, *to be affirmed* as the *affirmation* of *amor fati.*

In his 1969 *Nietzsche and the Vicious Circle,* Pierre Klossowski considers the "experience" of the eternal return, arguing that the discovery of the doctrine entails a peculiar kind of anamnesis and, with it, a necessary disruption of the identity of the one who discovers it. For in the discovery, I am conscious of myself as having discovered the recurrence of all things, and *with this* it must also be granted that the discovery itself has been made innumerable times before. If *I,* then, have learned the truth of recurrence an infinite number of times before, I must also have forgotten it infinitely many times before; the current anamnesis is possible only on the basis of this infinite prior forgetting. But the revelation, as that of the return of all things, is also the discovery that the revelation itself *could have happened* at any moment in the cycle, or to anyone in general, and has indeed happened innumerable times before. If, then, I *will* its repetition, I do not will the repetition of myself as I am now (having discovered the eternal recurrence) or as I was before discovering it. Rather, I will myself "as a fortuitous moment whose very fortuity implies the necessity of the integral return of the whole series," a series that must thus traverse "the successive realizations of all possible identities."[30] For this reason, at the moment I discover the eternal return

> I cease to be myself *hic et nunc* and am susceptible to becoming innumerable others, knowing that I shall forget this revelation once I am outside the memory of myself; this forgetting forms the object of my present willing; for such a forgetting would amount to a memory outside my own limits: and my present consciousness will be established only in the forgetting of my other possible identities.[31]

It is in this way that the circle becomes, according to Klossowski, "a sign for everything that has happened, for everything that is happening, and for everything that will ever happen in the world."[32]

With the discovery of the return, I discover what is on the one hand a general form of time and events, but what on the other implicates me, and the moment of discovery itself, as the outcome of the infinite sequence that also begins at this very moment. The particular moment of discovery is not itself unique; the discovery can happen at any time, and must indeed have happened an infinite number of times before. Accordingly, the possibility of self-disclosure that occurs in it—whereby *I,* in this moment, discover myself as the outcome of an infinite series of events that also begins in it—must be structurally characteristic of *every* moment as such. This possibility is in fact itself just a consequence of the essential possibility of the performance of an indexical recognition of one's own position within the world, or an indexical (self-)location of the present time as "now," which possibility characterizes, as such, every moment. This possibility is not, moreover, simply *my* possibility; in the revelation, I also understand that it is the possibility of anyone at any time. It must rather be seen as structurally implied in the form of the moment itself. Through this structure, every moment in itself communicates with the *whole* totality of infinitely recurrent time. The possibility of the revelation of this whole—the revelation of the eternal return, along with its correlative forgetting—must then be given in the very (paradoxical) form in which the moment gives itself. To determine my will in accordance with this form is then nothing other than to will the whole of everything that has happened, is happening, and will happen. This is to will in accordance with *amor fati.* The metaformal and ontological basis of the possibility of such a willing is given in the paradoxical form of given time—the kairological paradox of the moment—already and as such.

In this way and in both of its aspects—both in terms of what it suggests for the form of the "cosmic" time of the world and for the moment as such—the idea of the eternal return can be seen as *one* figure of the structure of given time as it is thought and experienced beyond the closure of the metaphysics of presence. There may be other such figures, and what is most important in the context of the development of the consequences of the ontological problematic of being and time here is, at any rate, not the "literal" or descriptive truth of the eternal return, but rather how it bears witness to the broader Real of infinite time as given outside this closure. At any rate, though, it is here that contact can be reestablished with Heidegger's own thought, in particular in that it construes the form of given time as a reflexive self-giving that discloses and conceals. *One* form of this self-giving is, as we have seen, the specific structure of Dasein as clearing and concealing, the structure that originally relates it to truth as *aletheia.* This is the structure in which, according to *Being and Time,* temporality temporalizes itself through the ecstases and in which Dasein is "its" time; behind it lies the reflexive structure of paradoxical auto-affection, or self-giving time, that Heidegger discovers through his reading of Kant in *Kant and the Problem of Metaphysics.* In this paradoxical structure, as we saw in chapters 4 and 6 above, the idea of transcendence is ultimately overcome in favor of the more ontologically basic clearing that precedes the attained Dasein itself and in favor of the epochal historical granting of being's truth as ground and abyss. But it remains that the specific structural condition for the revealing and concealing of beings and for the granting of being's truth is to be seen in what, giving itself, gives time.[33]

To interpret the implications of the ontological problematic this way is obviously to disagree with Heidegger's own interpretation of the significance of Nietzsche's doctrine and the closure of metaphysics it indicates. For instance, in the summer 1939 lecture course "The Will to Power as Knowledge," Heidegger portrays Nietzsche's position overall as an "inverted Platonism" in which is realized the final configuration of the succession of positions that Plato's own thought of being as beingness originally inaugurated.[34] It is such a position that actually underlies, according to Heidegger, the linked Nietzschean metaphysics of will to power and eternal return and explains their basic metaphysical orientation. This orientation, the "essence" of the will to power, is the inversion of the Platonic hierarchical opposition of being and becoming into a position that seeks the unlimited "permanentizing" of becoming in presence [*Beständingung des Werdens in die Anwesenheit*].[35] The idea of an *eternal* recurrence is that of a "securing" of the becoming of what becomes and a "permanentizing" of its "constancy in the direction of its circling back into itself and forward to itself [*im Sinne des in sich zurück- und zu sich vorauslaufenden Kreisens*]."[36] Such a "permanentizing" of becoming in the repeated form of the self-identical unfolds truth, in particular, as a "securing of beings in their decidable feasibility [*in seiner ausmachbaren Machbarkeit*]."[37]

As we have seen, though, in light of the development of the temporal paradoxes, the consequences of the temporal structure actually indicated here include an unlimitedness of *difference* in becoming, a paradoxical *undermining* of all fixed forms of identity, and above all a constitutive *undecidability* characteristic of the sense of the moment and of infinite given time as a whole. None of these consequences are thinkable under the headings of "stability," "permanentizing," and the identity of the same, with which Heidegger goes on to characterize Nietzsche's doctrine as the final position of metaphysical thinking. Together, though, they indicate a more complex and problematic relationship of the thought of the eternal recurrence, if construed in terms of them, to the determined history of metaphysics as a whole. In particular, they appear to bear to this history, as we have seen, the relationship that a general or dialectical problem bears to *one* of its more or less determined solutions.

What, though, is that which, in granting the form of time and events, predetermines the happening of whatever can happen in the determined history of metaphysics as the metaphysics of presence, indeed by predetermining the form of presencing and the present themselves? In Heidegger's own discourse, what does this is indicated by the name or descriptor *Ereignis*. And with the specific thought of these consequences of the eternal recurrence, thought in terms of the specific structures of the infinite indicated in the contemporary metaformal results, it thus apparently becomes possible to think that *the eternal recurrence is itself* one *form or structure of Ereignis*. In thinking this (apparently) un-Heideggerian thought, we will have also thought the possibility of the appearance of *Ereignis itself*, at a certain determined point of its discovery or revelation, in or as the infinite temporal form of the eternal recurrence. But inasmuch as the eternal recurrence is itself the metaphysical picture correspondent, according to Heidegger himself, to the contemporary regime of the dominance of machination or *Gestell*, to see things this way would simply be to see *Ereignis in Gestell* as its other face or initially hidden form. It is in this way that the moment can become, not an occasion for decision, but the object of an affirmation, an affirmation that goes all the further in implying its own infinite becoming. It is in this affirmation that time and being may be seen as showing themselves in a more original sense, and thereby as conditioning a life, beyond the closure of metaphysics, to come.

4. Non-Heideggerian Ontological-Critical Politics of the Present

If, as I have argued, the eternal return can be ontologically thought, in its determination of the very form of the happening of what happens, as "the same as" *Ereignis*, it is thus possible to think *Ereignis* quite differently than Heidegger himself does. This different thought of *Ereignis* also leads to quite a different thought of its consequences, and accordingly of a *present* which is evidently already determined by the forms of the completion and closure of the metaphysics which it also determines. What is at issue here, and to be brought into question on this basis, is also the evident ambiguity of Heidegger's own thought of *Ereignis* as the essentially *futural* event of being's own appropriation to itself, of its ultimate return to itself in the sphere of its own propriety or proper belonging, outside and beyond the withdrawal and obscuration within the history of metaphysics that is also, according to Heidegger, "proper" to it.

What, then, of the "event"ality of the event, of the thought or possibility of the determinate moment of a decisive break with all that has come before, or of the overcoming of the specific conditions of a longstanding historical or temporal determination of the sense of beings at a *particular* determined point? Many of the distinctive forms of temporal thought and political organization of the twentieth century (whether motivated in Marxist, fascist, or liberal-democratic terms) characteristically pursue a kind of sudden and radical breakthrough to the real at the punctual limit of the complete formalization of life.[38] In its rhetoric of the anticipation or production of an uncertain future, this pursuit recurrently invokes the peculiar ideological configuration formed by the figures of crisis, decision, and act. In these twentieth-century forms of organization and practice, the anxious anticipatory thinking of capitalist or technological crisis is thereby essentially linked to the ambivalent idea of an end of history, whether this is itself understood in catastrophic or triumphalist forms. But this configuration is already actually overcome by the thought of the temporal form of the eternal return as grounded in the *undecidability* of temporal sense. Here "ontological" thought is no longer responsive to an anxiety of the possible loss of what drives history or the exigency of its futural recovery on other terms or in terms of the "other." Rather, what is inherent and determinative in the events of history is thought in terms of *its own* specific structural and temporal relationship to what returns in it, and what thereby structurally implies its communication with all that happens, can happen, or will happen. With this thought of the infinite circulation of the undecidable, there is no longer anything to decide. There is no need, and no exigency, of the preservation, maintenance, or salvation of the event or its destiny, since this destiny also destines everything and everything returns. But there remains the possibility of an affirmation that determines the world and the will under its sign, thereby indicating the inherent temporal ground of the pursuit of a clarified life to come.

With this, what is figured in the figure of the eternal return becomes the principle of a life and action that is not determined by the total and effective production and manipulation of beings. This points to what is, within this life and action, a very different relationship to history and its becoming than that which is contemplated in the rhetoric of crisis and overcoming which capitalist-technological *praxis* and decisionist political thought essentially share. In particular, if Nietzsche, with the revelation of the eternal return, could consider himself in possession of a thought that would or could "break up the history of mankind," it is because it is part of the structural content of this thought that this specifically determined history is, as such and at each of its moments, *already in communication* with the temporality of what is not history or is other to it: that is, what cannot simply be determined as *what happens* according to history's determinate principles or determining forms of intelligibility.[39]

Here there is indicated, then, not the "other" beginning of "another history," but rather *other* and *different* histories or temporalities in general, and also the world-temporality of what is *not* history but is rather thought as the time of "nature" that precedes and envelops it. With this thought of the communication, in the paradoxical form of the moment, of history with its more general temporal conditions, history is *already* and as such "broken" with respect to its broader form in the circulation of the eternal return. It is this circulation of historical and natural temporalities, *in the plural*, and their communication within the paradoxical and infinite form of the now, that the eternal return then teaches.

To live self-consciously within this temporal condition is then to live without the hope of restoration or revelation, grasping that any disclosure to thought or understanding of the sense of the world would in fact decide nothing, since it would itself simply be another *undecidable* event. Here it is therefore possible to suspect or discern the specific structural limit of the great (and terrible) Heideggerian dramatics of being's concealment and revealing, and in particular of the massive narrative of being's progressive withdrawal within the history of metaphysics. In the thought of the eternal return as metaformally indicated here, by contrast, "nothing is hidden" and there is no withdrawal, unless it be the forgetting of what can and must always be remembered again, the circulation of what will always come back.[40] The result is that the thought that indicates and affirms the image of this circulation is no longer one that attempts to supply another or superior sense of being with respect to the sense of the circulation itself. The anxiety of nihilism, or its completion in the substitution of ontic for ontological measures, is then no longer threatening within an awareness that sees that, within the structural indication of the recurrence of all that happens, there is never too little meaning but always too much.[41] Under the sign of the eternal recurrence, that is, there is always *more* sense than can be grasped by means of finite procedures, always *more* truth than can be captured by resemblance or representation, always *more* time than we count in the span of our lives.

What kind of response, then, is invoked here *on the level of action* to the characteristic problems of our time, which present themselves as the problems of the *totalization* of the world, the pursuit of total enframing or the absolutization of the claim of technological means and practices over the circulation of life on the earth? Here we should think not only of the narrowly specified, though very real, problems of the exhaustion of resources and the verified and increasingly urgent disharmony of the global climate which push to crisis, but also of the internal "problems" of what Marx already discussed as the ongoing and ever more total circulation of life and energy in the planet's metabolism under the condition of capitalism, which could, for all we know, go on for a very long time to come.[42] If there is a response to these problems indicated by the ontological thought that is figured by the eternal return, it is not the suggestion of a restoration of a "natural" balance or a local or global amelioration of practices, still determined within the overarching form of capital and technological/effective solutions, which allow contemporary (human and nonhuman) ways of life to be maintained a bit longer. On the contrary, with respect to the totalization of the world, which these solutions will always themselves have both presupposed and promulgated, the affirmation of the eternal return is the affirmation of a circulation that can never be total at all, at least not within any overarching form of consistent presence.

The response (if such there be) is thus not to be understood on the level of a "solution" to the problem here posed, but rather in the pursuit of a *temporal life* that reflexively comprehends its more general and undecidable form. It is to understand, and find ways of living with, the time in which (as we recognize) what we do will have many consequences, but will never have *results* in the sense of *outcomes*: that nothing "comes out" because *nothing* is finally decided, and that to pursue a thereby clarified life is also to live *without* procedural or evental hope. To live in this form of temporality would evidently be to inhabit, *in an ontologically clarified way*, the present itself. This would not be to see it as the realization of a past future, or the potential of what is still "to come," but rather to inhabit, with understanding, the clarified structure of the time that also characterizes the being of the world, prior to and independently of us, and thus of all that *takes time* within it. In this respect, it would be to inhabit a life, a form of living being, that knows itself as the undecidable temporal form of life as such, or the unlimited sense of the earth. We are far from knowing what specific or concrete forms of action or practice such a life might require, or propose. But that the broad form of its temporal and ontological structure can nevertheless be preliminarily indicated, under contemporary conditions, appears to be among the most poignant implications of the thought of being today.

Notes

Preface

1. An interpretation of the ontological implications of the infinite as it is understood in contemporary mathematics is attempted, along partly similar lines, by Alain Badiou in *Being and Event.*

2. Livingston, *The Politics of Logic,* chap. 1.

3. By Mark Wrathall; see Wrathall, *Heidegger and Unconcealment,* 41 (from an article first published in 1999).

4. "Ontology" and "ontological" are used throughout this work in reference to the questions of the sense and truth of being, rather than (as in some contemporary discussions) to the project of establishing a catalogue of entities or their categories in answer to the question "What exists?"

5. I have addressed, in a broader way, what I see as some of the specifically political consequences of logic and formalism in *The Politics of Logic,* esp. chap. 10.

6. For this reason, in the online appendix 1, "Heidegger, Truth, and Time," I have included some supporting material that was cut from an earlier version of the main text and does not figure directly in the main line of argument as thus construed. This includes, in particular, (i) exegetical passages commenting more or less directly on Heidegger's texts; (ii) considerations of, and responses to, existing secondary literature on Heidegger; and (iii) clarifications of the relationship of the current project to Heidegger's own stated views and commitments. Readers who are interested in following up on any of these issues are invited to consult this appendix. A second appendix, "Time, Number, and Ideal Genesis in Aristotle and Plato," also culled from an earlier version, develops a phenomenological and ontological interpretation of the problem of the relationship of number and time as it figures in Plato and Aristotle's metaphysical thought and in dialogue with twentieth-century interpreters of Heidegger. Both appendixes are available online at https://www.unm.edu/~pmliving/.

Chapter 1

1. Plato, *Sophist,* 246a.
2. Ibid., 264a–b.
3. Ibid., 246b–c.
4. Ibid., 246e–247a.

5. Ibid., 247c–e.

6. This problem is also closely related to—or perhaps identical with—the final and "great" difficulty pointed out by Parmenides himself, in discussion with the young Socrates, for the doctrine of timeless forms which the latter attempts to defend in the first part of the *Parmenides* (133b).

7. "*to gar auto noein estin te kai einai*" (Diels and Kranz, Fr. 3; quoted in Clement, *Stromateis*, 6.23, and Plotinus, 5.1.8); both the translation and the textual provenance of the remark are controversial. See Kirk, Raven, and Schofield, *The Pre-Socratic Philosophers*, 246.

8. Plato, *Sophist*, 237a–e; for statements of closely related "Parmenidean" problems about saying, knowing, or judging what is not, see also Plato, *Theaetetus*, 188e–189b, and *Cratylus*, 429c–e.

9. Plato, *Sophist*, 254e–255a.

10. Ibid., 257b.

11. Ibid., 250b–d.

12. Plato, *Sophist*, 253a. Compare Wittgenstein, *Philosophical Investigations* (henceforth *PI*), sec. 371: "*Essence* is expressed in grammar."

13. Plato, *Sophist*, 253c.

14. Ibid., 253c–d.

15. Ibid., 256d.

16. Ibid., 256e. Quotations in this sentence and previous one taken from Fowler translation.

17. Plato's formulation here closely anticipates the famous "definition" of truth and falsehood that Aristotle gives at *Metaphysics* 4.1011b, in the course of a defense of the universality of the principle of noncontradiction: "To say of what is that it is not, or of what is not that it is, is false, while to say of what is that it is, and of what is not that it is not, is true."

18. GA 2, 1.

19. Plato, *Sophist*, 242c.

20. GA 19, 441.

21. The ontological difference is defined as such in the summer 1927 course "Basic Problems of Phenomenology" (GA 24, 22–23): "We must be able to bring out clearly the difference between being and beings in order to make something like being the theme of inquiry. This distinction is not arbitrary; rather, it is the one by which the theme of ontology and thus of philosophy itself is first of all attained. . . . We call it the *ontological difference*—the differentiation between being and beings. Only by making this distinction—*krinein* in Greek—not between one being and another being but between being and beings do we first enter the field of philosophical research. Only by taking this critical stance do we keep our own standing inside the field of philosophy."

22. GA 19, 446.

23. Ibid., 453.

24. Ibid., 446–47.

25. Ibid., 486.

26. Ibid., 487 (transl. slightly modified).

27. Ibid., 552.

28. Ibid., 466–67 (transl. slightly modified; emphasis in Heidegger's original).
29. Ibid., 342 (transl. slightly modified).
30. See chap. 2.
31. GA 19, 495.
32. See, e.g., *Politics* 1253a7–10 and *Politics* 1332b5 (though neither passage literally contains the phrase in question).
33. GA 2, 22.
34. Ibid., 25.
35. Ibid., 14.
36. Aristotle, *De Anima* 3.8, 431b21; GA 2, 14.
37. GA 65, 8.
38. The connection of metaphysics with the thought of the totality of beings is already explicit, before the "turn," in Heidegger's 1929 Freiburg inaugural address, "What Is Metaphysics?" (GA 9), and develops considerations about "worldview" and totality that go back much earlier, at least to the 1919 "War Emergency Semester" course "The Idea of Philosophy and the Problem of Worldview" (GA 56/57). Compare, also, "The Age of the World-Picture," from 1938, in *Off the Beaten Track* (GA 5).
39. GA 11, 68. Compare Iain Thomson's perceptive analysis of the "twofold" operation of grounding that is characteristic of onto-theo-logy for Heidegger: "All successful, epoch-grounding metaphysical systems *combine* these two different forms [ontological and theological] of foundationalism, thereby securing our understanding of the being of entities (and so grounding the intelligible order) from *both* the inside out *and* the outside in, microscopically *and* telescopically, floor to ceiling—or, as Heidegger puts it, ontologically *and* theologically, that is, *ontotheologically*" (Thomson, *Heidegger on Ontotheology*, 18–19).
40. GA 11, 65 (transl. slightly altered).
41. Ibid., 59.
42. Ibid., 76.
43. Ibid., 67.
44. Frege, *Grundgesetze*, xvii–xviii.
45. Frege, "Logic," 228.
46. Ibid., 227.
47. Frege, "Thought," 326.
48. Frege, "Logic," 228; the parallel passage in "Thought" is 326–27.
49. Cf. Dummett, "Can Truth Be Defined?," 443–45.
50. E.g., in "Thought," Frege says: "It is also worth noticing that the sentence 'I smell the scent of violets' has just the same content as the sentence 'It is true that I smell the scent of violets'" (328).
51. Of course, Frege's own concerns about truth were much more directed toward questions about the content of thought rather than about (linguistic) sentences.
52. See the essays collected in Davidson, *Inquiries into Truth and Interpretation*, esp. "Truth and Meaning," "In Defence of Convention T," and "Radical Interpretation."
53. Davidson, *Truth and Predication*, 27–28.

54. Dummett, "Truth."

55. Davidson, *Truth and Predication*, chaps. 1 and 2; cf. Davidson, "The Folly of Trying to Define Truth."

56. Quoted and cited in Martin, *Theories of Judgment*, 75.

57. "A judgment will always be expressed by means of the symbol ⊢, which stands at the left of the symbol or complex of symbols which gives the content of the judgment. If the small vertical stroke at the end of the horizontal one is *omitted*, then the judgment will be transformed into a *mere complex of ideas*, of which the writer does not state whether he recognizes its truth or not." Frege, *Begriffsschrift*, 52.

58. Frege, "Function and Concept," 142.

59. On the ambiguity involved in any treatment of the judgment stroke as having logical significance (for instance the significance of a predication of facthood or of truth) see Martin, *Theories of Judgment*, chap. 3.

60. At other places (see, e.g., "My Basic Logical Insights," 323–24) Frege gestures toward the inarticulate relationship between logic and force that shows up, not indeed in the sense or reference of any linguistic term by itself, but in the peculiar *failure* of the predicate "true" to contribute positively to the sense of sentences in which it figures. Frege argues that the unique structural role that "true" plays (in that, specifically, it adds nothing to the sense of *any* sentence in which it figures) with respect to *all* the sentences of a language that allows it to "[seem] to" manifest the specific force of assertion that is involved in any predication whatsoever to appear *as* an element of a predicative sentence. But the attempt "miscarries"; it is not, in fact, possible for a predicate signifying this involvement to appear significantly as a predicate within a sentence. Nevertheless, in the miscarriage itself, "what is characteristic of logic" is formally and negatively "indicated." Cf. also Martin's trenchant discussion of the "miscarriage" of the truth-predicate as pointing to necessary features of logic in relation to truth that bear close comparison with what Heidegger formulates as the implications of the "ambiguity of the copula" in *Theories of Judgment.* See also Dummett, *Frege and Other Philosophers*, 247–48.

61. "'Being' cannot be derived from higher concepts by definition, nor can it be presented through lower ones. But does this imply that 'Being' no longer offers a problem? Not at all. We can infer only that 'Being' cannot have the character of an entity. Thus we cannot apply to Being the concept of 'definition' as presented in traditional logic, which itself has its foundations in ancient ontology and which, within certain limits, provides a justifiable way of characterizing 'entities.' The indefinability of Being does not eliminate the question of its meaning; it demands that we look that question in the face" (GA 2, 4).

62. Frege, "On *Sinn* and *Bedeutung*," 151. *Erkenntniswert*: standard translations render this as "cognitive value," but this is somewhat misleading, given Frege's thoroughgoing separation of sense from anything "cognitive" in the sense of "psychological."

63. These devices were independently discovered, in an essentially similar form, by Peirce.

64. I pass over here the very significant develops of algebraic logic in the

nineteenth century beyond the Aristotelian picture, many of which may have significantly influenced Frege himself.

65. Frege, "On Concept and Object," 181–93.

66. Ibid., 184–85.

67. And thus, *a fortiori* (as we shall see), a difference that cannot be captured as a "type" of "form" capable of mixing with others, as in the *Sophist.*

68. Cf. Dummett, *The Interpretation of Frege's Philosophy*, 541: "The very sharpness of Frege's distinction between objects and concepts makes it impossible to compare his doctrines at all fruitfully with those actually advanced by Plato."

69. More technically: Basic Law V says that any two concepts, F and G, have the same value-range if and only if the objects that are Fs are exactly those that are Gs. Under some interpretations, the law itself can actually be separated from the assumption of universal comprehension which is its background in Frege's system and it has been argued that the law can be upheld as an axiom schema, in the context of more limited comprehension principles (particularly ones that prohibit impredicativity), without producing inconsistencies of the Russellian type. For discussion and an argument to this effect, see Heck, "Predicative Fragments." But see also Incurvati and Murzi, "Maximally Consistent Sets," for some essential complications for this kind of suggestion.

70. Frege, "Thought."

71. GA 21, 170–82.

72. Ibid., 183–84.

73. Ibid., 191–92.

74. Ibid., 192–93.

75. Ibid., 168.

76. Ibid., 168.

77. Ibid., 161.

78. Ibid., 194–95.

79. Ibid., 194–98.

80. GA 20, 8.

81. Ibid., 8.

82. Ibid., 2, 6.

83. GA 26, 38.

84. Ibid., 40–42.

85. Ibid., 44.

86. Ibid., 112. This suggestion bears interesting parallels to the thesis and argument of Russell's 1900 book on Leibniz (Russell, *A Critical Exposition of the Philosophy of Leibniz*), according to which Leibniz draws his metaphysics of the monad and the preestablished harmony essentially from his subject-predicate logic.

87. GA 26, 85.

88. Ibid., 141–42. Cf. Heidegger's *Introduction to Metaphysics* (GA 40).

89. GA 26, 199 (transl. slightly modified). As Jon Cogburn has pointed out to me, Heidegger's concern here is closely related to some of those that motivate the early Schelling.

90. Ibid., 270.

91. Ibid.

92. Ibid., 271.

93. See also Meillassoux, *After Finitude*, for a consideration of some of the possible implications of the mathematical infinite for realism about time.

Chapter 2

1. GA 14, 98.

2. Husserl, *Logical Investigations*, 589–92, 594.

3. Ibid., 477. This is Husserl's phenomenological formulation of what Heidegger repeatedly cites, in the Latin, as the motto of the traditional "correspondence" theory of truth.

4. Ibid., 600.

5. Ibid., 601.

6. Ibid.

7. Ibid., 478.

8. Ibid., 634.

9. Husserl thus does not here yet distinguish categorial intuition from what he will later call "eidetic" intuition or the showing of essences (including material ones); for eidetic intuition, see, e.g., Husserl, *Ideas* and *Experience and Judgement.*

10. Husserl, *Formal and Transcendental Logic*, 47–48, 57–58. Husserl uses Aristotle's term "apophansis" to characterize (as Aristotle does) the specific phenomenon of a declarative statement or judgment (*logos*) and "apophantic" to designate something having the character of this phenomenon.

11. Husserl, *Formal and Transcendental Logic*, 134.

12. Ibid., 138.

13. Ibid.

14. Both dissertations are included in GA 1.

15. See, in particular, "Die Lehre vom Urteil," part 5, chap. 2, secs. 4–6 (GA 1, 118–29), and "Die Kategorien- und Bedeutungslehre des Duns Scotus" (GA 1, esp. 276–83). For discussions of the influence of Lask on Heidegger's thinking in these years, see van Buren, *The Young Heidegger*, esp. chaps. 4 and 11; Kisiel, *Genesis*, esp. chap. 1; Kisiel, "Why Students of Heidegger Will Have to Read Emil Lask"; and Crowell, *Husserl, Heidegger*, chaps. 2 and 3.

16. GA 56/57, 41.

17. Ibid., 60.

18. Ibid., 59–60.

19. Ibid., 75.

20. Ibid., 63.

21. Ibid., 116.

22. Ibid., 117.

23. GA 63, 3.

24. Ibid., 25–26.

25. Ibid., 7, 21.

26. Ibid., 29.

27. Ibid., 21, 25–26.

28. Ibid., 29. Heidegger adds: "cf. intentionality and its *arche.*"
29. Ibid., 69.
30. Ibid., 73–74.
31. Ibid., 72.
32. Ibid., 79–80.
33. Ibid., 80.
34. GA 60. A number of commentators have discussed the methodology of formal indication and its employment in *Being and Time.* See, e.g., Kisiel, *Genesis;* Dahlstrom, "Heidegger's Method"; van Buren, "The Ethics of Formale Anzeige"; Crowell, *Husserl, Heidegger;* Dreyfus, "Comments on Cristina Lafont's Interpretation of *Being and Time*"; Lafont, "Replies"; and Shockey, "What's Formal about Formal Indication?"
35. Husserl, *Ideas,* 26–27.
36. Ibid., 27.
37. Ibid., 28.
38. Kisiel, *Genesis,* 165–70.
39. Formalization in this sense might thus almost be defined as "structuralization," provided the sense of "structure" is ontological or logical rather than stipulative or conventional, and provided that structural location is understood as having the essentially demonstrative character emphasized here.
40. Formalization in this sense also bears some evident connections to the sense in which Kant characterizes both space and time as "forms" of intuition in essential connection with the definitive capacity of intuition to present a particular. However, Kant's picture of givenness, as depending crucially on the distinction and relation between the forms of the intuition and the conceptual and categorical forms of the understanding, is itself open to interrogation on the basis of a more extended ontological and temporal application of formal indication; see chap. 6 below.
41. GA 61, 13–36.
42. Ibid., 16–17.
43. Ibid.
44. Ibid., 17.
45. Ibid. (Heidegger gives the formulation in Latin).
46. Ibid., 18.
47. Ibid., 20 (transl. slightly modified).
48. Ibid.
49. Ibid., 33.
50. Ibid., 34.
51. Ibid.
52. Ibid.
53. Ibid.
54. Ibid., 35.
55. Nevertheless, this is no reason (as we shall see) to conclude that the results cannot be articulated and shared in public and intersubjective language.
56. For this reason, as well, the formal indication is not a matter of demonstrating *conditions of possibility* for a phenomenon but rather of indicating its

actual features; as such and for this reason, formal indication must be sharply distinguished from any project of "transcendental" reflection in the narrowly Kantian sense.

57. As such a demonstration, the formal indication is evidently analogous in some ways to the methodology of "showing" suggested by Wittgenstein in the *Tractatus* for the demonstration of logical form. However, there is no evident reason why what is demonstrated in a formal indication must be "unsayable," incommunicable, or "interior"; and for this reason as well, formal indication does not involve the idea of "private" ostensive definition critically targeted in the *Philosophical Investigations.*

58. See Shockey, "What's Formal about Formal Indication?," for a clear presentation of this feature as well as some of the others. And for more discussion of the particular sense of "grounding" that is involved here, see chap. 3.

59. GA 63, 29.

60. Van Buren, *The Young Heidegger,* 283.

61. Kisiel, *Genesis,* 423.

62. GA 63, 29.

63. Ibid., 30.

64. Ibid., 31.

65. Ibid.

66. Ibid.

67. GA 2, secs. 25–27, 34–35.

68. "All *explicata* to which the analytic of Dasein gives rise are obtained by considering Dasein's existence-structure. Because Dasein's characters of Being are defined in terms of existentiality, we call them '*existentialia.*' They are to be sharply distinguished from what we call '*categories*'—characteristics of Being for entities whose character is not that of Dasein" (GA 2, 44).

69. Both the need for such a methodology, and the benefit of the demonstrations it produces, are palpable against the backdrop of reflection on (what can then be seen) as the failures of analytic approaches to the analysis of the same phenomena. Beginning with Frege's problematic argument, in "Thought," for a privileged sense of the self by means of which each subject is presented to him- or herself, analytic methodologies have recurrently evinced the problem of characterizing the logical form of indexicality in relation to an objectively describable structure of semantic relationships characteristic of the meaningfulness of (objective) language overall. For the by-now-familiar problem of the "essential indexical" which appears not to be capturable in terms of any objective description but nevertheless capable of bearing epistemic content and import for a subject, see (e.g.) Perry, "The Problem of the Essential Indexical."

70. See, e.g., Brandom, *Making It Explicit* and *Articulating Reasons.*

71. Brandom, *Making It Explicit,* 384.

72. For a critique along these lines, see Livingston, *Philosophy and the Vision of Language,* chap. 8. See also the discussion of the paradox of the institution of language in chap. 4 below.

73. For "conceptual role" semantics, see, e.g., Sellars, "Empiricism and the Philosophy of Mind"; for the roots of the idea of "topic-neutral" analysis see, e.g.,

Smart, "Sensations and Brain Processes"; Lewis, "An Argument for the Identity Theory"; and Lewis, "Psychophysical and Theoretical Identifications."

74. For this problem, see, e.g., Chalmers, *The Conscious Mind*, see also the discussion of the underlying form of the problem in Livingston, *Philosophical History and the Problem of Consciousness*, chaps. 1 and 6.

75. See, e.g., Cavell, *Must We Mean What We Say?*, esp. chaps. 1 and 4, and *The Claim of Reason*, esp. chaps. 4, 6, and 7.

76. This methodology is perhaps best exemplified, and its closeness to the methodology of formal indication shown, by Wittgenstein's insistence on the provision of "reminders" of what we—in some sense—already know, and is thus already present "in" our language, insofar as we live and speak it; this is a methodological point that is not necessarily captured as fully, or at any rate is not brought out as explicitly, in practitioners such as Ryle and Austin. For more discussion, see chap. 6, section 2.

77. For example, Cavell often presents his project as one of bringing back an element of the "human," specifically the "human voice" to philosophy, over against the dominance of the formal-symbolic approaches characteristic of analytic philosophy during the 1950s, '60s, and '70s. However, in order for the relationship of the methodology of ordinary language philosophy he defends to be clarified in its relationship to that of formal indication, it would first be necessary to clarify just what is meant by the "human" here—especially in view of the "Heideggerian" question of the provenance of the *definition* of the human as the *zoon logon echon* and its tendency to "personalistic" formulation (see sec. 2 above)—as well as (of course) the senses in which, and the reasons why, the aims and methods of this kind of project are indeed in irreducible tension with those of formal-symbolic formalization.

78. For some discussion of Heidegger's own rejection of formal-symbolic methods, see the online appendix 1, "Formalism and the Question of Being" (supplemental to chap. 5).

79. GA 20, 36.

80. Ibid., 48–49.

81. Ibid., 37–39.

82. Ibid., 69.

83. Ibid., 69–72.

84. Ibid., 71.

85. Ibid.

86. Ibid.

87. Ibid. (transl. slightly modified).

88. Ibid., 52 (transl. slightly modified).

89. GA 20, 140.

90. Ibid., 139.

91. Ibid., 155–56.

92. Ibid., 174.

93. "Dasein ist an ihm selbst von Hause aus welt-offen, offen für die Welt, die ihrerseits aufgeschlossen ist" (GA 21, 143).

94. Ibid., 164 (transl. modified).

95. The first horn of this dilemma is discussed in more detail in chaps. 3 and 6 below; the second is familiarly posed by Sellars in "Empiricism and the Philosophy of Mind." See also McDowell, *Mind and World,* for a clear presentation of the basic dilemma here.

96. GA 20, 99–100.

97. Ibid., 101.

98. Ibid.

99. Ibid.

100. Ibid., 102.

101. Ibid.

Chapter 3

1. In particular, I do not wish to inscribe the implication of a definitive connection, either of motivation or result, to Kant's transcendental idealism or any subsequent (e.g., "fundamental ontological") version of it. As we shall see, it is essential to the conception of truth to be extracted from the unified reading of the Davidsonian and Heideggerian programs that it be able to be construed in a completely *realist* way.

2. Davidson's argument against correspondence is a version of the argument that has been called the "slingshot" and which is given in (somewhat different) classic forms by Church and Gödel. For discussion and further references, see Davidson, *Truth and Predication,* 38–41, 126–30. For Heidegger's differently motivated argument, see his discussion in *Being and Time* (GA 2, 214–18), which emphasizes the direct phenomenal presentation of phenomena, independent of any need for a representational doubling of them. Both arguments bear comparison to Frege's argument against correspondence truth (discussed above in chap. 1). Some commentators, e.g., Wrathall (*Heidegger and Unconcealment,* 12–13) and Carman (*Heidegger's Analytic,* 159–61), have read Heidegger as holding that correspondence theories actually provide an accurate account of propositional truth itself, in that an assertion can indeed be considered to be true just when it corresponds with a state of affairs it is "about," provided this propositional truth is seen (as usual) as a limited phenomenon within the broader horizon of truth as unconcealment. (I take no position on whether or not this is correct as a reading of Heidegger.)

3. For Davidson's arguments against antirealist views, see Davidson, *Truth and Predication,* 44–48. For Heidegger's, see, e.g., GA 24, 315–16, where Heidegger holds that truth is a presupposition for "our own existence" rather than conversely, as well as GA 2 (*Being and Time*), 227–28, for a similar (but more briefly stated) claim.

4. In Heidegger's case, this rejection is motivated by his critique of Husserl's failure to pose the question of the *ontological basis* of the realm of "ideal contents" (see chap. 2). In Davidson's case, it is motivated largely by his inheritance of Quine's devastating arguments against the intelligibility of any such notion of contents or meanings.

5. For Davidson's program of the provision of a "theory of meaning" through the radical interpretation of natural language, see, e.g., "Truth and Meaning" (1967); "Radical Interpretation" (1973); and "In Defense of Convention T" (1973), all reprinted in Davidson, *Inquiries into Truth and Interpretation.* Following standard practice in the "analytic" literature, I here use "natural language" to indicate a contrast with "formal" or "artificial" languages, and thus actually to refer to what Heidegger, by contrast, generally calls "historical" languages.

6. Tarski, "The Semantic Conception of Truth."

7. Tarski, "The Concept of Truth in Formalized Languages."

8. See, e.g., Davidson, "Truth and Meaning," 17, and "Radical Interpretation," 125–26.

9. Davidson, *Truth and Predication,* 16–17.

10. Ibid., 27–28.

11. Ibid., 28.

12. Ibid., 31.

13. Ibid., 10–14. Despite possible anticipations in Frege (see chap. 1 above), what many have seen as the first explicit suggestion of a "redundancy" theory is given by F. P. Ramsey in 1927 in "Facts and Propositions."

14. Davidson, *Truth and Predication,* 26–27.

15. Heidegger uses "unconcealment [*Unverborgenheit*]" through much of his career (especially after 1928) as a maximally general term for the discussion of the phenomenon of truth; as such it appears to be intended as a translation or near-translation of the Greek term *aletheia,* which also exhibits the privative structure of a lack or failure of concealment or obscurity. In *Being and Time* itself, though, Heidegger barely uses "unconcealment"; instead he typically uses "uncoveredness [*Entdecktheit*]," both as a synonym for "unconcealment" and also more narrowly, in reference to entities (especially as they are shown or uncovered in assertions), while "disclosedness [*Erschlossenheit*]" is used in relation to Dasein or its constitutive phenomenon of world. I am grateful to Mark Wrathall for pointing out some of these distinctions to me in conversation; see also Wrathall, *Heidegger and Unconcealment,* 12–17.

16. GA 2, 218–19.

17. Ibid., 220.

18. In this sense, in particular, "assertion and its structure . . . are founded upon interpretation and its structure" (GA 2, 223).

19. Ibid., 158.

20. Ibid., 157. Many commentators have emphasized the implications of this primacy of circumspective unconcealment and the mode of the *zuhanden,* especially in the way it undermines theoretical or Cartesian assumptions about the primacy of cognition or representation in our relation to entities. See, e.g., Dreyfus, *Being-in-the-World*; Harman, *Tool-Being*; Okrent, *Heidegger's Pragmatism*; and Braver, *Groundless Grounds,* for some diverse developments of this theme.

21. Ibid., 158.

22. Ibid., 150.

23. Ibid.

24. Ibid., 151.

25. Ibid.

26. Davidson, *Truth and Predication*, chaps. 4–5.

27. Ibid., 159–61.

28. Ibid., 161.

29. GA 24, 311–12.

30. Dahlstrom, *Heidegger's Concept of Truth*, 17.

31. Tugendhat, "Heidegger's Idea of Truth" and *Der Wahrheitsbegriff.*

32. Tugendhat, *Der Wahrheitsbegriff*, 344–45.

33. Ibid., 342.

34. Ibid., 345.

35. Tugendhat here repeats the problem posed by Parmenides, and answered by Plato's Eleatic Visitor, for judgments of or about nonbeings (see chap. 1 for discussion). For the clearest development of Heidegger's own alternative response to the question of the possibility of error and falsehood, whereby it is itself grounded in the threefold "fore-"structure of the existential-hermeneutic "as," see GA 21, 187–90.

36. Tugendhat's critique of Heidegger's concept of truth has been widely discussed in secondary literature on Heidegger, wherein many possible responses have also been assayed. See, e.g., Wrathall, "Heidegger and Truth as Correspondence"; Wrathall, *Heidegger and Unconcealment*, 34–39; Carman, *Heidegger's Analytic*, 259–61; and Dahlstrom, *Heidegger's Concept of Truth*, 398–409. For a critical response upholding (in part) Tugendhat's critique, see Lafont, *Heidegger, Language, and World-Disclosure*, 115–24. For a helpful overview of the whole discussion also partially upholding the trenchancy of Tugendhat's critique, see William H. Smith, "Why Tugendhat's Critique of Heidegger's Concept of Truth Remains a Critical Problem." For more on Tugendhat's critique and responses to it, see the online appendix 1, "Tugendhat's Critique of Heidegger's Concept of Truth" (supplemental to chap. 4).

37. See sec. 1 of this chapter.

38. GA 2, 68–69.

39. Ibid., 149–50.

40. Ibid., 34.

41. To say that the two kinds of grounding are *compatible* in this sense, though, is not to say that they are *consistent*, either with each other or each with itself. Indeed, as I shall argue in chaps. 4 and 5 below, the twofold grounding developed here leads, indeed as a result of its close relation to the ontological difference, to the metalogical dilemma of incompleteness (if consistency is maintained) or inconsistency (if completeness is asserted).

42. In the "preparatory analytic" of *Being and Time* (esp. division 1), this "across" is thought primarily in terms of the structure of Dasein, as marked by the ontic-ontological "priority" that is definitive of it: namely, that of a being that has, in its being, a constitutive concern with being. As Heidegger would often suggest later, though, what is actually primary here is not Dasein itself but rather the ontic-ontological structure definitive of it; or (put another way) it is not the activity or experience of Dasein, but rather the ontic-ontological structure that conditions and grounds *its* (factical, ontic) existence that is most decisive here.

For this reason, Heidegger will later conceive of Dasein itself as conditioned by the deeper structure of the "open" that is granted by being, or *Ereignis*, and of this "open" in or as which being—or *Ereignis*—conditions a domain of presencing as the deeper structure of (being's) truth. For further discussion, see chap. 4.

43. Here, as we have seen, "presence" irreducibly has both a temporal and a (quasi-)spatial meaning: for this reason, such an account of ontological grounding can ultimately (as we shall see in chap. 6) provide a unitary ontological form for the theorization of both *sense* and *given time.*

44. We should, again, not think of this as a matter of being instituting, creating, or mandating the *conditions of possibility* either for the existence of beings or for the (subjective) experience of them. For "opening" and "domain" here indicate not only the conditioning of beings, but *also* of their intelligibility or sense. As Heidegger suggests, in the judgment *that A is B* there is cojudged the *possible* being-B of A, or in other words that A has the *sense* of possibly being (a) B. But it is this *sense* that is ontologically grounded in the first instance, and the possible ways that A can be are subsequently grounded in *it*: it is only so far as A is *intelligible as* (something that can be a) B that A *can be* presented or uncovered as B. The possibilities of entities are thus ontologically grounded in their projective sense, rather than this sense being grounded in the possibilities. As we shall see in more detail in chap. 4, this projective sense is itself grounded (under the condition of the history of metaphysics) in the determinate epochs of intelligibility and the kinds of opening they permit, and these are *themselves* (further) grounded in the "truth" of being across the ontological difference (as *Ereignis*).

45. The *locus classicus* of the argument is, of course, Frege, "On *Sinn* and *Bedeutung*."

46. For the figure of the "third realm," see especially Frege, "Thought."

47. Frege, "On *Sinn* and *Bedeutung*," 153–54.

48. Carnap, *Meaning and Necessity*, secs. 28–32.

49. Russell appears to have been the first to point this out (in "On Denoting").

50. Carnap, *Meaning and Necessity*, 8, 142–44.

51. Ibid., 8.

52. See Hanzel, "Frege, the Identity of Sinn, and Carnap's Intension," 243–44, for a version of this point.

53. Quine, "Truth by Convention" and "Two Dogmas of Empiricism."

54. It should be noted that Carnap recognizes the problem that interchangeability according to logical rules is not sufficient to establish synonymy (*Meaning and Necessity*, secs. 12–14). But to answer it, he proposes a further criterion of sentence synonymy (sec. 14) which is itself to be defined in terms of the behavioral dispositions of speakers, and which should not (for familiar reasons) be seen as unproblematic in a post-Quinean context.

55. Cf. Livingston, *Philosophy and the Vision of Language*, chap. 5, and Livingston, *The Politics of Logic*, chap. 2. Essentially the same problem has also been discussed, in the more recently popular context of possible world semantics, as that of "hyperintensionality."

56. Dummett, *Frege: Philosophy of Language*, 227.
57. Dummett, *The Interpretation of Frege's Philosophy*, 132.
58. Dummett, *Frege: Philosophy of Language*, 227.
59. Ibid., 229.
60. Ibid.
61. Ibid., 231–32.
62. Dummett, *The Interpretation of Frege's Philosophy*, 132.
63. Dummett, *Frege: Philosophy of Language*, 233.

64. There are other precedents for generalizing Frege's notion of modes of presentation in this way. In 1969, Dagfinn Føllesdal argued that Husserl's notion of noema or noematic sense can be seen as a generalization of Frege's notion of sense as mode of presentation (Føllesdal, "Husserl's Notion of Noema"). On the suggested generalization, the noematic sense is that ideal component or aspect of an intentional act by means of which it achieves an intentional connection to a particular object or objectivity; the generalization results from relaxing Frege's restriction of the notion to the case of language and allowing it to characterize the structure of intentional acts of consciousness generally (thus Husserl can speak of the particular *noemata* of acts of perception, memory, imagination, etc.). For another suggestive recent development of the extended implications of Føllesdal's suggestion, bringing Husserl's conception of phenomenological presentation explicitly into connection with Tarski's understanding of truth, see David W. Smith, "Truth and Epoché."

65. This is not to suggest that the argument so far is sufficient to illuminate how they *should* then be treated; only that an adequate accounting for the phenomena of the informativeness of identity judgments which leads Frege to invoke the sense/reference distinction, along with the related phenomena of presentation we have considered, will *not* be possible simply in terms of the (ontic) relations of entities without also considering their (ontological) sense.

66. Dummett again recognizes the general issue here: "If I am wrong . . . and there is some non-circular account of the notion of knowing what it is for something to be the case, not construed as verbalizable knowledge (on pain of circularity in the course of explaining what it is to understand words), nor appealing to the recognition of its being the case, then the general form of the model of sense we have been considering can be preserved while the whole model is recast in terms of this notion. What we are interested in . . . is, after all . . . the structure of a model of sense, and not the question how far epistemological notions can legitimately be employed within such a model. But, despite the fact that Frege was undoubtedly highly realist in his whole philosophical outlook, that for him sense was related to truth rather than to the recognition of truth, and despite his constant inveighing against the intrusion of psychological notions into logic . . . it is far from apparent that he would have rejected an account of the form we have been considering, on the grounds of the present objection [viz., that the model in terms of recognition 'invokes epistemological considerations where they are relevant']" (Dummett, *The Interpretation of Frege's Philosophy*, 239–40). He goes on to argue that the notion of sense for

Frege is, after all, connected to "cognitive notions" involved in the advance from a thought to a truth-value, and that this suggests that it is wrong to hold that "Frege wanted to extrude everything epistemological from logic or from theory of meaning." But while it is true that the idea of "cognitive value" to which Frege appeals in introducing the concept of sense is, at least partly, an epistemic *notion*, though, this does not suffice to establish that Frege would be willing to accept any general epistemic *constraint*, of the sort Dummett imagines, on the nature of sense.

67. This temporality is further developed in chap. 4.

68. Tarski, "The Semantic Conception of Truth."

69. Kripke, "Outline of a Theory of Truth."

70. Gupta and Belnap, *The Revision Theory of Truth*, 113–18.

71. Ibid., 118–19.

72. Priest, *In Contradiction*, 12.

73. Ibid., 135.

74. Ibid., 21 (Priest explicitly addresses earlier versions of the revision theory given by Gupta alone and by Herzberger).

75. Ibid., 21–22.

76. Ibid., 22–23.

77. "As the disclosedness of the 'there,' understanding always pertains to the whole of Being-in-the-world. In every understanding of the world, existence is understood with it, and *vice versa* . . . [The] circle of understanding is not an orbit in which any random kind of knowledge may move; it is the expression of the existential *fore-structure* of Dasein itself . . . In the circle is hidden a positive possibility of the most primordial kind of knowing" (GA 2, 152–53).

78. Ibid., 152.

79. Ibid., 153, 315.

80. For discussion of one related way in which a certain idea of passage to the infinite and (Cantorian) transfinite can itself be motivated by a phenomenological consideration of the reflexivity of Dasein, without nevertheless disputing Dasein's essential structural finitude, see Becker, *Mathematische Existenz*, and appendix 2 to this work.

81. It is plausible that this kind of duality already appears as a direct consequence of Russell's paradox and the related limit-paradoxes that establish (given certain auxiliary assumptions) that the assumption of a total set or a set of all sets is inconsistent. Both Priest and Gupta and Belnap mention the close formal connections between the issues of consistency and completeness that arise in considering the T-structure and the set-theoretical paradoxes. For more discussion of these connections, see also Livingston, *The Politics of Logic*, chap. 1.

82. As we shall see in more detail in chap. 5, the "both"/"and" structure suggested here with respect to circularity and inconsistency gains a more determinate structure in the context of Heidegger's consideration of the distinction between the various epochal determinations of the being of beings within the history of metaphysics, on the one hand, and the "truth" of being as thought outside (or beyond the closure of) this history, on the other.

Chapter 4

1. For "world-entry," see "On the Essence of Ground" (GA 9), 158–59. See also discussion in chap. 1 above.

2. It is also important to distinguish the resulting *undecidability*—in this sense and as I will use it throughout this work—from indeterminacy, underdetermination, or truth-value gappiness. That a phenomenon is "undecidable" in the current sense means only that the values of its specific instances cannot be wholly and consistently determined by means of finite procedures, *not* that its value is unreal, illusory, or indeterminate.

3. Heidegger's "Seyn," which he uses throughout the *Beiträge* to indicate being as it might be thought outside the closure of metaphysics, is here translated as "beyng."

4. GA 65, 6 (trans. slightly modified).

5. Ibid., 76, 7.

6. Ibid., 7.

7. Ibid. (transl. slightly modified).

8. Ibid., 11.

9. Ibid., 10–11 (transl. slightly modified).

10. Ibid., 7.

11. GA 5, 204/221; also quoted in Schürmann, *Heidegger on Being and Acting,* 230.

12. GA 11, 73.

13. Cf. *Beiträge* (GA 65), secs. 36–38, 281.

14. *Vom Wesen der Sprache* (GA 85); cf. also *Logik als die Frage nach dem Wesen der Sprache* (GA 38), esp. secs. 6, 29, and *Sein und Wahrheit* (GA 36/37)

15. Cf. Livingston, *The Politics of Logic,* chap. 1.

16. See, e.g., "On the Way to Language" (GA 12), 186/175. See also Livingston, *Philosophy and the Vision of Language,* chap. 7 for discussion.

17. GA 12, 185/174.

18. Schürmann, *Heidegger on Being and Acting,* 17.

19. Schürmann, *Broken Hegemonies,* 96–97.

20. Cf. Wittgenstein, *PI,* §50: "One would like to say, however, that being cannot be attributed to an element [*Man kann dem Element nicht Sein beilegen*], for if it did not *exist,* one could not even name it, and so one could state nothing at all about it [*und also gar nichts von ihm aussagen*].—But let us consider an analogous case. There is *one* thing of which one can state neither that it is 1 metre long, nor that it is not 1 metre long, and that is the standard metre in Paris." (Wittgenstein's claim, that the Paris meter stick cannot be called either one meter long nor not one meter long, has sometimes been criticized, e.g., by Kripke in *Naming and Necessity.* The issue does not affect the present point, as far as I can tell.)

21. Davidson, "On the Very Idea," 183.

22. Davidson, "Truth and Meaning," 27, and "Belief and the Basis of Meaning," 153.

23. Davidson, "Belief and the Basis of Meaning," 153.

24. Davidson, "On the Very Idea," 197.

25. Ibid., 191.
26. Ibid., 192.
27. Ibid., 193–94.
28. Ibid., 189.
29. Ibid., 190, quoting Kuhn, "Reflections on My Critics," 266–67.
30. See, e.g., *Vom Wesen der Wahrheit: Zu Platons Höhlengleichnis und Theätet* (GA 34), esp. sec. 9 and following.
31. This is not to say that the hegemonic terms of earlier epochs (for instance Descartes's *res cogitans*) must always be identified as having a single, specific referent in newer ones, but only that the earlier organizing terms must, in general, be *understandable* (as they were earlier used and intended) from the perspective of the later one.
32. In *The Structure of Scientific Revolutions*, for example, Kuhn suggests that at a moment of crisis, the shift from one paradigm to the next is largely or fully determined by such factors as the surrounding political climate, the availability of a cohort of younger researchers not already convinced by the existing paradigm, or the overall organization of the "community life" of scientists. See, e.g., Kuhn, *Structure*, 92–96.
33. See chap. 3, sec. 4, above.
34. Davidson himself makes this point in "Theories of Meaning and Learnable Languages" (1965).
35. Although *epochs* and the economies of thought, language, and practice that characterize them are obviously not themselves formal systems, the considerations which motivate the *application* to them of results about formal systems and their properties (including undecidability) are thus essentially similar to those that motivate Davidson's own project of developing formal "theories of meaning" for natural languages. For some discussion of these considerations, see Davidson, "Theories of Meaning" and "Semantics for Natural Languages"; Dummett, "What Is a Theory of Meaning? (I)." This application can also be seen as motivated in a "Heideggerian" context, along essentially similar lines, by issues surrounding the form of "human" finitude in relation to the infinity of meaning or sense; see Livingston, "The Sense of Finitude," for this argument.
36. This structural undecidability is closely related to the phenomenon of the "undecidable" which Jacques Derrida has often discussed and understands as essential to deconstructive practice. See Livingston, "Derrida and Formal Logic," for a discussion of the formal structure of the undecidable, in Derrida's sense.
37. It might be objected that the undecidability of sense mentioned in the paragraph above does not sustain any conclusion this strong about the structural constitution of languages themselves, since what is (really) undecidable is not languages themselves but just their translation (for instance, *given* an interpretation of a language's sentences, there is a trivial decision procedure for deciding its T-sentences: one just checks, given a translation t[P] of a sentence P, whether the relevant T-sentence is of the form: t[P] is true iff P). (I owe this objection to Jon Cogburn.) However, since there is no sharp criterion for identity or difference of a language over time, what is at issue in the structural transformation of a language over time is just whether and how it can be "translated" into its

own earlier form; and for this reason the undecidability of translation is both intimately related to the "indeterminacy" of translation (which characterizes all natural languages, according to Quine and Davidson) and central to the question of linguistic structure itself.

38. GA 9, 131–33.

39. Ibid., 133–34.

40. Ibid., 158–59: "'Dasein transcends' means: in the essence of its being it is *world-forming* [*weltbildend*], 'forming' [*bildend*] in the multiple sense that it lets world occur, and through the world gives itself an original view [*Anblick*] (picture [*Bild*]) that is not explicitly grasped, yet functions precisely as a paradigmatic form [*Vor-bild*] for all manifest beings [*alles offenbare Seiende*], among which each respective Dasein itself belongs [*das jeweilige Dasein selbst gehört*].

Beings, such as nature in the broadest sense, could in no way become manifest unless they found *occasion* to enter into a world. This is why we speak of their possible and occasional *entry into world.* Entry into world is not some process that transpires in those beings that enter it [*kein Vorgang am eingehenden Seienden*], but is something that 'happens' 'with' beings [*sondern etwas, das 'mit' dem Seienden 'geschieht'*]" (transl. slightly modified).

41. GA 5, 33/30–31 (transl. slightly modified).

42. Ibid., 37/35.

43. Ibid., 35/32.

44. Ibid., 38/36.

45. Ibid., 39/37.

46. Ibid., 41/39.

47. Ibid., 41/39–40.

48. See, e.g., Rorty, "Heidegger, Contingency, and Pragmatism."

49. Cf., e.g., *Beiträge* (GA 65), 98–99, in the course of a discussion of the "decision" regarding "history" as the "playing out of the strife between earth and world": "What prepares itself, then? The *transition to the technologized animal* [*Übergang zum technisierten Tier*], the one that, through the gigantism of technology, is beginning to replace the instincts, which are already becoming weaker and coarser.

For this direction of decision what is characteristic is not the technologizing of 'culture' and the imposition of a 'worldview,' but is the fact that 'culture' and 'worldview' become resources of a battlefield technology for the sake of a will that no longer wills any goal; for the preservation of a people is never a possible goal but is only a condition of the setting of a goal . . . What then disappears completely is the possibility of knowing that 'culture' and 'worldview' are already the scions of a world-order which presumably is to be overcome [*bereits Ableger einer Weltordnung sind, die angeblich überwunden werden soll*]."

50. The idea of such a critical ontological analysis that operates as a clarifying critique of logical laws is often suggested by Heidegger and is a key component of his more general critical consideration of the "logical" tradition and the history it determines. But although Heidegger carried out detailed critical ontological analyses of such general and traditionally regulative logical-ontological principles as the principle of sufficient reason and the "law of identity," there

is no parallel sustained investigation anywhere in his corpus of the principle of noncontradiction. For the first, see GA 10 (1955–56); for the second, see "The Principle of Identity" in GA 79 (1957). It should be noted that there is a lecture given by Heidegger in Freiburger Kränzchen on 16 December 1932, entitled "Der Satz vom Widerspruch." As the lecture has not yet been published, however, I have been unable to consult it for this project.

51. Priest, *Doubt Truth to Be a Liar,* chap. 1. The law is stated at *Metaphysics* 4 (Γ), 1005b19–20, as the principle that "the same attribute cannot at the same time belong and not belong to the same subject in the same respect."

52. Grim ("What Is a Contradiction?," 51–54), for example, gives some nineteen different and nonequivalent statements of the nature of contradictions from classical and contemporary literature, distinguishable (among other differences) according to whether they treat it as a matter of contradictory propositions, logical forms, assertions and denials, or states of affairs.

53. In introducing the terminology, I do not wish to invite possible confusion with what Jeff Malpas, in an insightful analysis of the role of place in Heidegger's thought, terms Heidegger's own "topology" of being (Malpas, *Heidegger's Topology,* 27–37). Malpas draws the term itself in part from a late retrospective remark of Heidegger's, from the Le Thor seminar in 1969 (GA 15, 344), in which he appears to indicate a transition from the earlier questions of the meaning and truth of being to a final question concerning the "place or location [*Ortschaft*] of being."

54. Because it is essentially a picture of the space of contents rather than of the operation (or truth-function) of negation itself, this is not directly equivalent to any of the various types of logical negation that have been discussed in recent literature. However, the sense of negation that most directly supports the topological picture is the "classical" one that allows both double negation elimination and contraposition, and generally supports argument by *reductio.*

55. For an illuminating discussion of the (very different) uses of the principle in Spinoza, Kant, and Hegel, see Melamed, "Omnis determinatio est negatio." Compare also Deleuze's vehement rejection of the principle, in the context of a conception of the positivity of difference and its priority to negation, in Deleuze, *Difference and Repetition,* 52–53.

56. Wittgenstein, *Tractatus Logico-Philosophicus,* 2.06, 2.063, 4.462. Cf. also 4.0641.

57. As Andrew Cutrofello has pointed out to me, a more complete analysis than can be carried out here would have to consider also the conception of "infinite" or "indefinite" judgment, which Kant (possibly following Aristotle) distinguishes from negative judgment (in the context of a term logic rather than a sentential logic) on the ground that, while negative judgment affirms the negation of a predicate ("The soul is not mortal"), infinite judgment simply locates the subject in the infinite space *outside* a given predicate ("The soul is nonmortal"). Ultimately, however, "indefinite" judgment appears still to fall broadly within the topo-logical picture as I describe it here, since it is still deployed (at least according to Kant) within what is effectively a picture of the *complete* and *noncontradictory* possible determination of judgments by means of it (see Longuenesse, *Kant and*

the Capacity to Judge, 295–96, for discussion of this "principle of complete determination" in Kant).

58. What, then, of the "contradictions" that underlie the dialectic for Hegel? Though I do not take a position here on the (vexed) question of the right way to understand the Hegelian *Aufhebung*, it does seem clear that, insofar as contrary terms are there seen as recurrently generating higher unities, their seeming or apparent contradictions are to be resolved or (in some sense) dissipated at the level of a higher totality.

59. See chap. 1.

60. At *De Interpretatione* 16a12–13, in the course of a discussion of the nature of the *logos* or sentence, Aristotle says (somewhat cryptically) that "falsity and truth have to do with combination and separation." *One* possible interpretation of this and the remarks following it is that whereas an affirmative judgment combines what is said by the subject term with what is said by the predicate term, a negative judgment separates them. However, this interpretation does not appear to be established definitively on the basis of Aristotle's text. For some discussion of the issue, see the opening pages of Husserl's *Experience and Judgement*, and compare also Heidegger (GA 2, 159).

61. "Impredicative" definitions are those that appeal to a totality of entities that includes what is to be defined; impredicative terms are those that can only be thus defined.

62. See Beaney, "Introduction," 19–20, for a discussion of the general issue.

63. In fact, as has often been pointed out, this is already implicit in Cantor's theorem, which he proved using a variety of the general method known as "diagonalization" (cf. Beaney, "Introduction," 20; Grim, "There Is No Set of All Truths").

64. Priest, *Beyond the Limits of Thought*, 151–53.

65. See chap. 3.

66. Priest, *Beyond the Limits of Thought*, 153.

67. Ibid., 124.

68. Not necessarily, as we have seen, only in cognition, representation, or linguistic expression, but also possibly in "nonlinguistic" contexts of action, practice, and practical engagement.

69. Priest, *Beyond the Limits of Thought*, chap. 15. In *One* (2014) and "The Answer to the Question of Being" (2015), Priest has subsequently argued for answering Heidegger's question of the being of beings with a theory of unifying, though contradictory, "gluons" that unify individual entities as well as beings as a whole.

70. This conclusion is sometimes put—though, I think, misleadingly—as the conclusion that there simply "is no" such totality. We will revisit the position that, in view of the paradoxes, maintains overall consistency by denying the existence of a *complete* world of entities in chap. 5 below.

71. This has the evident consequence that, if Heidegger is indeed faithful to his own guiding question, his distinction between presence at hand [*Vorhandensein*] and readiness to hand [*Zuhandensein*] cannot (contra the recent suggestion of McDaniel, "Ways of Being") simply be read in terms of irreducibly distinct quantifiers or "quantifier variance."

72. As it is announced and maintained in the tradition since Parmenides, the law thus wavers in an overdetermined fashion between prohibition and the pointing out of what is anyway supposed to be impossible; for a discussion of this structure as it is exhibited both in Parmenides and in the early Wittgenstein, see Livingston, *The Politics of Logic*, chap. 5.

73. Cf. chap. 3, sec. 3

Chapter 5

1. Gödel, "Some Basic Theorems," 134.

2. This terminology is not meant to indicate any philosophically tendentious view of the "thing in itself" or the "in-itself" as a particularly determined region or structure. It simply means "that which is" as such.

3. This repudiation may be seen as finding expression, for example, in the strong antihumanist rhetoric of Heidegger's "Letter on Humanism" of 1946–47 (GA 9).

4. Lacan, *Encore*, 93.

5. Lacan's concept of the "Real" is complex and undergoes many changes of specification and inflection over the course of his career. I do not take a view here about how precisely to define it or which formulation is most important, but seek only to preserve the link that is constitutive for Lacan between the Real and formalization at the latter's point of inherent impasse. For a very exhaustive and illuminating treatment of Lacan's concept, see Eyers, *Lacan and the Concept of the "Real."* I also discuss Lacan's motto and Badiou's reversal of it into his own claim for a "theory of the pass of the real, in the breach opened up by formalization" in Livingston, *The Politics of Logic*, 188–92.

6. For the definition, see Turing, "On Computable Numbers."

7. A bit of care needs to be taken here. By "formal system" I mean here any system that is *axiomatizable* or effectively enumerable. The reason for restricting the discussion to "formal systems" in this sense is that it is just these to which considerations about effectivity and effective procedures apply. As I have suggested in chapter 4, as well as in Livingston, "The Sense of Finitude," it is in terms of the structure of such systems that considerations about the finitude of the representation of the structure of languages by their users are naturally posed.

8. I here state the first theorem in a somewhat rough and intuitive version. For discussion of the issues involved in the difference between this and other possible statements, see Livingston, *The Politics of Logic*, chap. 6. See also Berto, *There's Something about Gödel*, chap. 2, for more detailed discussion.

9. For a system to "capture the capacities of human mathematical thought" would be (in the relevant sense) for it to be able to deduce just those theorems we are capable (in some sense) of proving. Of course, even stating things this way requires a significant (and in many ways problematic) idealization of our "capability."

10. Gödel, "Some Basic Theorems," 133.

11. Ibid., 130.

12. The result that Gödel refers to in 1951 is that the consistency statement is equivalent to some statement of the form $\forall x_1 \ldots x_n \, \exists y_1 \ldots y_m \, [p(x_1, \ldots, x_n, y_1, \ldots, y_m) = 0]$ where p is a polynomial with integer coefficients and the variables range over natural numbers; later the work of Davis, Putnam, Robinson, and Matiyasevich showed that one can replace the statement with something of the form $\forall x_1 \ldots x_n \, [p(x_1, \ldots, x_n) \neq 0]$, For discussion, see Feferman, "Are There Absolutely Unsolvable Problems?," 6. I am indebted to John Bova for first pointing out to me the significance of the Davis et al. results for questions of realism and metalogical difference.

13. I follow here the trenchant exegesis of Gödel's conclusion in the Gibbs lecture given by Feferman, "Are There Absolutely Unsolvable Problems?," 1–7.

14. I here develop the conception of *orientations of thought* suggested by Alain Badiou in his 1998 *Briefings on Existence*:

I call an "orientation in thought" that which regulates the assertions of existence in this thought. An orientation in thought is either what formally authorizes the inscription of an existential quantifier at the head of a formula, which lays out the properties a region of Being is assumed to have. Or it is what ontologically sets up the universe of the pure presentation of the thinkable. (53)

For a more detailed exposition of the four orientations of thought which I distinguish (adding one, the paradoxico-critical, to the three described by Badiou in 1998), see Livingston, *The Politics of Logic*, 51–60.

15. This reference is thus at least compatible with the "characterization principle" suggested by Routley (*Exploring Meinong's Jungle*) and Priest (*Towards Non-Being*), according to which every object (existent or not) has the properties it is characterized as having (at least when these are not "existence-entailing"). For this reason it is also compatible with the "noneism" they defend on its basis, although I do not think it *requires* noneism or neo-Meinongianism as an answer to questions about existence and quantification. Furthermore, the framework of possible, impossible, and "open" worlds to which Priest appeals in *Towards Non-Being* would itself raise questions, in the present context, about the basis for *its* total availability.

16. Does this definition make the idea of an "incomplete" totality—which I consider below as a commitment of the generic orientation in particular—a contradiction in terms? It does not. For while adherents to the generic orientation often put their position as claiming that "there is no One-All" or involving simply the nonexistence of the "whole" in question here, the underlying claim (in terms of the framework of the orientations) is actually not about existence but about thought: it is that thought always fails to *comprehend* the totality (of whatever is).

17. As we have seen (chap. 4), even Kant's position in this sense invokes a complete and consistent whole, in positing the totality of things in themselves as the *thinkable* (although not knowable) basis of appearances and applying to them the principle of noncontradiction.

18. Cf. Livingston, *The Politics of Logic*, chaps. 1 and 8.

19. See also Bova, "Negation and Incommensurability" and *A Metalogical Approach to Platonic Dialectic*, in which he develops the implications of metalogical difference for negation, for the Socratic *elenchos*, and for Sartre's conceptions of the in-itself and for-itself.

20. As we saw at the end of the last chapter, the basis for this realism about the ontological difference is just the formulation of the ontological question itself. It can be denied, but only if the question of being itself is repressed or ignored.

21. For some such considerations, bearing in favor of the paradoxico-critical orientation and against the generic one, see Livingston, *The Politics of Logic*, chaps. 9 and 10.

22. Cf. Badiou's discussion of the "axiomatic of contemporary belief" at the opening of *Logics of Worlds*.

23. As above, I restrict the discussion here to "theories" that are axiomatizable.

24. To put things this way, in terms of Gödel's results about formal systems strong enough to express arithmetic, is obviously to put them somewhat differently from the statements above in terms of the totality of *what is*, to which results such as Russell's paradox or Cantor's theorem may seem to have a more direct and obvious bearing. However, many have noted the suggestive parallels between Russell's paradox and Gödel's results, and in particular the similarity in the way that both seem to indicate the necessity of a broad and general decision about the power of thought in relation to total reality: i.e., the "forced choice" between the consistent thought of an incomplete totality and the inconsistent thought of the whole. (For some discussion of this, see Livingston, *The Politics of Logic*, chap. 1.) As Jon Cogburn has recently pointed out to me, this connection is more closely confirmed through recent results by Incurvati and Murzi, extending earlier results by McGee about truth ("Maximal Consistent Sets"). Incurvati and Murzi show that the restriction of comprehension principles, in the wake of Russell's paradox, to include only "maximally consistent" sets of instances of comprehension: (i) yields many such sets of instances, which are not mutually compatible; and (ii) yields sets of instances which (if "maximally" consistent in a natural sense) are not effectively decidable. Since the second result is an application of Gödel's theorems, this shows specifically how the failure of the attempt to maintain consistency jointly with (decidable) completeness may be seen in terms of effectivity and Gödel's results.

25. Gödel himself, particularly in his later years, was, as is well known, a dedicated antimechanist, and sometimes referred to his incompleteness theorems as providing evidence against mechanism. However, in his more careful statements, such as those in the Gibbs lecture, he is always careful to affirm at most the disjunctive conclusion, at least one disjunct of which, as we have seen, appears consistent with mechanism about the mind. See Shapiro, "Incompleteness, Mechanism, and Optimism," for an overview of the (large) literature on Gödel and mechanism, as well as the conclusion that "there is no plausible mechanist thesis on offer that is sufficiently precise to be undermined by the incompleteness theorems" (275). See also Berto, *There's Something about Gödel*, chap. 11, for citations and discussion.

26. I thus follow Feferman, "Are There Absolutely Unsolvable Problems?," 11, in considering that, even if there are problems with applying Gödel's reasoning directly to the question of mechanism, "at an informal, non-mathematical, more every-day level, there is nevertheless something to the ideas involved [in

his argument for the "disjunctive conclusion"] and something to the argument that we can and should take seriously."

27. In *The Politics of Logic* (291), I called this position simply "formal realism." I add the prefix "meta-," here, to reflect that what is concerned is not primarily an attitude (e.g., a Platonist one) about the "reality" or "actual existence" of forms, but rather the implications of the *transit* of forms in relation to what is thinkable of the real, the transit that can, in view of Cantor's framework, be carried out beyond the finite.

28. It is true, as Andrew Cutrofello has pointed out, that Gödel himself sometimes puts his realism in terms that suggest an onto-theological position (one invoking, for instance, a divine absolute intellect capable of knowing what human thought cannot). However, as we shall see, the real upshot of the undecidability shown in his incompleteness results as well as the broader phenomenon of the "inexhaustibility of mathematics" to which they bear witness is to suggest a "mathematical reality" which cannot be comprehended *even* by an intellect supposed to possess infinitary (though determinate) powers.

29. Gödel says this about the second term of the disjunction: "The second alternative, where there exist absolutely undecidable mathematical propositions, seems to disprove the view, that mathematics (in any sense) is only our own creation . . . So this alternative seems to imply that mathematical objects and facts or at least *something* in them exist objectively and independently of our mental acts and decisions, i.e. to say some form or other of Platonism or 'Realism' as to the mathematical objects" (Gödel, "Some Basic Theorems," 135–36).

30. Nor is it any help to posit, in onto-theological fashion, an absolute divine intellect endowed with infinitary powers of thought and thus (as we might suppose) capable of knowing all the mathematical truths (from a self-consistent standpoint) even if *we* cannot. In "Transfinite Recursive Progressions of Axiomatic Theories," Feferman has shown that there is indeed a kind of "completeness" of arithmetic truths that is obtainable by the transfinitely repeated application of so-called "reflection principles," each of which amounts to adopting as a new axiom for a new system certain assumptions about the consistency of an earlier system (or the truth of its results). By means of an appropriate transfinite procedure through these principles, it is indeed possible, as Feferman shows, to obtain the totality of arithmetic truths. However, this "procedure" is itself not specifiable in a recursively enumerable way, so it does not provide anything like a general effective procedure for determining arithmetic truth. See Shapiro, "Incompleteness," and Berto, *There's Something about Gödel*, chap. 11, for discussion.

31. As a consequence of this, *both* the paradoxico-critical and the generic orientations are essentially committed to (because they result from an attitude of) metaformal realism. This introduces an apparent lack of symmetry among the four orientations, in that constructivism is then the only of the four which is (systematically or globally) antirealist. However, as I argue in "Metaformal Realism and the Ontological Problematic" (Colorado College talk), the symmetry can be restored by means of the consideration that constructivism and the generic orientation distinguish, where the other two do not, between *global* and *local* applications of logical principles, and thereby provide grounds for further distinc-

tions in how realism and antirealism are to be applied in each case. Specifically, whereas constructivism is *globally* (transcendentally) antirealist but *locally* (empirically) realist (compare Kant's arguments for the conjunction of transcendental idealism and empirical realism), the generic orientation is, by contrast, *globally* realist but *locally* (with respect to logically constituted "worlds") antirealist. By contrast with both, both the onto-theological and the paradoxico-critical orientations are realist on both the global and local levels.

32. Dummett, "Realism" (1963); for some later (1978) reflections on the development of the framework and issues related to it, see Dummett, "Preface."

33. Dummett, "Preface," xxx–xxxii.

34. See, e.g., Dummett, "Realism," 146.

35. In this situation, the consideration that justifies the label "realism" is the same as that in the noncontradictory case: that, by stark contrast with the antirealist option, the truths (including possibly the contradictory ones) are seen as rigorously and exhaustively determined, quite independently of considerations of knowledge, practice, or use.

36. Dummett, "The Philosophical Basis."

37. Ibid., 216–17.

38. Ibid., 225.

39. Ibid., 247.

40. Dummett, "Preface," xxxii–xxxvix, and "The Reality of the Past."

41. Dummett, "The Philosophical Significance," 186.

42. Ibid.

43. Ibid., 187.

44. Ibid., 190.

45. Ibid., 192.

46. Ibid., 195.

47. This is just an application of the Domain Principle (discussed in chap. 4 above).

48. In appendix 2 to the current work, developing the suggestions of Becker in *Mathematische Existenz* and Lautman, "New Research," I argue for a way of understanding how numbers and the transfinite may be understood as "given" and constituted on the basis of ontological difference, even (and especially) in light of Gödel's transformative 1931 incompleteness results.

49. In particular, even if the incompleteness results are seen as indicating the existence of *problems* forever unsolvable, their very objectivity as problems turns on the consideration that their possible answers nevertheless have determinate truth-values.

50. This is importantly different from the claim that the law of excluded middle must be maintained *internally* to the particular logical system in question. There are, for instance, "intuitionistically valid" demonstrations of Russell's contradiction. That is, there is an "intuitionistically valid" proof, not employing the law of excluded middle, of the conclusion that $(R \in R) \wedge \sim(R \in R)$, where R is the Russell set. However, if this result is interpreted (more broadly) in the antirealist *meta*-theoretical *setting* wherein it is taken to establish or demand the "nonexistence" of R, it is no *paradox*. For *on this interpretation*, even the

"intuitionistically valid" argument has a false premise, namely that R exists, and thus, since unsound, does not suffice to establish its conclusion. Somewhat similarly, Gödel's first theorem can itself be proven using only constructivist methods, but this does not imply that, in the absence of a metatheoretical application of bivalence (i.e., the claim that each sentence is either true or false), the demonstration is itself capable of indicating the (actual) metaformal (or ontological) situation.

51. Taylor Carman (*Heidegger's Analytic*, 155–58) reads Heidegger as propounding an "ontic realism" but also holds that on the *ontological* level Heidegger is neither realist nor idealist and that he is there engaged in a "transcendental enterprise" involving making explicit the "hermeneutic conditions that render . . . entities intelligible to us" (155, 158). As we have seen, the position I am here calling "ontological" realism should be distinguished from ontic realism, in that it is not an attitude toward entities at all and is compatible with various such attitudes. On the other hand, it is essential to it that the way that it demonstrates the dependence of the intelligibility and sense of entities on *Dasein* (or, later, the "clearing") does not extend to showing them to be dependent on "us" *as* (for instance) empirically-anthropologically described, but rather only on the ontic-ontological structure of Dasein as such.

It is worth noting that, while ontological realism as thus understood is not primarily a "metaphysical" realism, it also may be reasonably be thought to capture from a more metaphysically neutral perspective some aspects of many of the (various) formulations of metaphysical realism. For example, since ontological realism is also as such a realism about the subject matter of statements about the *totality* of entities, it might be seen as capturing the formally legitimate core of metaphors such as that of a "view from nowhere" or "god's-eye perspective" on entities as a whole, or the possibility of a *complete* description of them. Somewhat similarly, metaformal realism underwrites realism about the *totality* of T-sentences for a language—including those, as we saw in chap. 3, which lead to paradox—and thus might reasonably be thought to capture the formal core of metaphysically realist "correspondence" intuitions about truth, although (as we also saw in chap. 3) it does not require or support anything like an actual "correspondence" theory.

52. "Statements" should here be understood as "things said," without prejudice (at least in the first instance) to their status as tokens or types.

Chapter 6

1. GA 3, 28.
2. Ibid., 31–32.
3. Ibid., 31.
4. Ibid., 37–39.
5. Ibid., 38.
6. Ibid., 32.
7. Ibid., 33.

8. Ibid., 34; cf. Kant, *Critique of Pure Reason* (henceforth *KRV*), A 68–69 / B 93–94.

9. GA 3, 34.

10. Ibid.

11. Ibid.

12. *KRV*, A 78 / B 103.

13. Ibid.

14. Cf. *KRV*, A 106–7.

15. GA 3, 77; cf. *KRV*, A 118.

16. *KRV*, A 118.

17. GA 3, 79 (trans. slightly modified); cf. *KRV*, A 98–99.

18. GA 3, 86.

19. Ibid., 87.

20. *KRV*, A 138 / B 177.

21. *KRV*, A 140 / B 179.

22. Ibid.

23. *KRV*, A 140 / B 179–80.

24. *KRV*, A 141 / B 180.

25. *KRV*, A 141–42 / B 180–81.

26. GA 3, 93–94.

27. GA 3, 99; cf. *KRV* A 145 / B 184 and A 138 / B 177.

28. GA 3, 101. Cf. *KRV* A 144 / B 183: "The schema of substance is the persistence of the real in time, i.e., the representation of the real as a substratum of empirical time-determination in general, which therefore endures while everything else changes."

29. GA 3, 101. This despite the fact that, as Heidegger recognizes (GA 3, 97), Kant says very clearly, a page or two earlier, that the schemata of *pure* concepts of the understanding, such as substance, "can never be brought into any image whatsoever" (*KRV*, A 143–44 / B 183).

30. GA 3, 101.

31. Ibid., 102.

32. Ibid.

33. Ibid.

34. Ibid.

35. Ibid., 172.

36. Ibid.; *KRV*, B 67.

37. GA 3, 172; *KRV*, B 67–68.

38. *KRV*, B 68–69.

39. "Pure self-affection provides the transcendental, primal structure of the finite self as such. Thus it is absolutely not the case that a mind exists among others which, for it, are also something related to it, and that it practices self-positing. Rather this 'from-out-of-itself-toward . . . and back-to-itself' [*'Von-sich-aus-hin-zu . . . und Zurück-auf-sich'*] first constitutes the mental character of the mind as a finite self" (GA 3, 173).

40. Ibid., 140.

41. Ibid., 177. It is also notable that essentially the same paradoxical struc-

ture is explicit in Husserl and implied by his analysis of the structure of internal time-consciousness as an ultimately foundational form of synthesis. See, e.g., Husserl, *Cartesian Meditations*, 43.

42. GA 3, 173.

43. Ibid., 174–75.

44. Ibid., 167; *KRV*, A 152 / B 192.

45. He envisions, in particular, replacing statements of the form "No A is B and not-B at the same time" with something of the form "No non-B A is B," which eliminates the need for a temporal determination and thus renders it (according to Kant) analytic.

46. GA 3, 177.

47. Compare the title and methodology of Heidegger's second "being-historical" monograph of the 1930s, *Besinnung* (GA 66).

48. For an interpretation of the connection between the rule-following considerations and the private language argument along these lines, see Kripke, *Wittgenstein on Rules and Private Language*, 1–7, 60–62. For a related interpretation of the connection between the targets of the two skeins of argument in relation to Heidegger's own critique of the contemporary configuration of "machination [*Machenschaft*]" and "lived experience [*Erlebnis*]," see Livingston, "Thinking and Being."

49. Cf. also *PI* 186.

50. Wittgenstein's challenge to the "logic of identity" here, as it figures in the spontaneous ideology of rule-following, is closely related to Heidegger's own ontologically grounded critique of the traditional "law of identity" A = A in "The Principle of Identity" (GA 79); see also Livingston, "Thinking and Being."

51. As Andrew Cutrofello has pointed out to me, Kant addresses a closely related (or perhaps the same) problem at A133 / B 172, where he invokes in response a "transcendental power of judgment" which, as distinct from the general logic, "cannot be taught but only practiced" and requires for its possession a "natural gift," in the absence of which the correct application of rules cannot be guaranteed.

52. Each of these features is, arguably at least, decisively present both in Kant's account in the *Critique of Pure Reason* and Heidegger's in *Being and Time.* In bringing them out in this way, I am thus not primarily rejecting the account of temporality given in either book, but rather simply further formalizing the structure which can (arguably at least) be seen as common to both, once they are purged of the ideas of constitutive transcendental subjectivity (Kant) or a quasi-subjective temporally constituting Dasein (Heidegger). For more on the relationship of this formalization to Heidegger's own account of temporality in *Being and Time*, see appendix 1.

53. GA 2, secs. 65–68.

54. GA 3, secs. 43–45.

55. For this reason, according to Kant, "The infinitude of time signifies nothing more than that every determinate magnitude of time is only possible through limitations of a single time grounding it" (A 32 / B 47–48). In this sense, according to Kant, the "original representation" of time must be "given as unlim-

ited," but can only be so given in an "immediate intuition" and cannot be thus represented or determined by means of concepts.

56. I here employ again Iain Thomson's perceptive formulation of the twofold mode in which every onto-theo-logical figure constitutes the sense of beings as such and as a whole (see Thomson, *Heidegger on Ontotheology*, 18–19, and chap. 1 above). As we shall see, this characteristic twofoldness can be inherited, in a post-metaphysical condition, by understanding each as relating to a distinct way in which time itself is given.

57. A 414–17 / B 441–45.

58. Although the topic of time appears officially only in the first antinomy, in parallel with the question of the boundedness of space, the question of time is in fact central to all four antinomies. For all four turn on the question of the totality of series of conditions, and this question in each case raises the issue of the structure of *priority* thought more or less explicitly in temporal terms as the question of conditioning, either on the side of the subject or the object or both.

59. Priest, *Beyond the Limits of Thought*, 88–89, 96–101. For the structure of inclosure paradoxes, see chap. 4 above.

60. Priest, *Beyond the Limits of Thought*, 96–98.

61. The structure and statement of this paradox is closely related to (and in part derived from) the paradoxical structure of the divided "now" which Derrida locates in Aristotle's treatment of time in the *Physics* in "Ousia and Gramme" (1968). For a recent and trenchant treatment of the structure and its implications for ethics and religion, see Hägglund, *Radical Atheism*, chaps. 1 and 2, and for another helpful analysis, in connection with Aristotle, see Brogan, *Heidegger and Aristotle*, esp. 74–77. For more discussion of connections to Aristotle as well as Plato, see also the online appendix 2 to this work.

62. Of course, given relativity, there is no simultaneity in an "absolute" sense—that is, in a sense that is not relative to particular inertial reference frames. Hence it might be thought problematic to speak of the totality of what is "now" happening in this sense. However, the argument of this paragraph can be read as implicitly relativized to a particular reference frame and thus as having a conclusion bearing on the becoming of the world as *seen from that reference frame*; as far as I can tell, everything goes through as before.

63. More explicitly, for an inclosure paradox that fits Priest's schema (*Beyond the Limits of Thought*, 134): let $\Omega = \{y; \varphi(y)\}$ where φ (y) says that y is happening *now*. Let $\psi(x)$ be: x is changing. Then Ω (the present state of the world, considered as the totality of all that is now happening) exists, and ψ (Ω) is satisfied, since the world is now changing. Finally, for an arbitrary subset x such that x is changing, let $\delta(x)$ be the circumstance that something in x is becoming something else—that is, that some entity a in x which is some way b is also *becoming* not-b (or not fully b).

Then we have Transcendence: $\delta(x)$ is not an element of x. That is because, for any a whose *being-b* is in x, its *becoming* not-b is sufficient to make it the case that x is (as a whole) changing and is thus not included in x. But we also have Closure: $(\delta(x) \in \Omega)$. For this, its becoming, is indeed already happening *now*. This is sufficient to produce the contradiction that the becoming of the present (or of the world as a whole) both is and is not happening in the present.

64. This is, as we have seen, compatible with Kant's claim that time is "empirically" real (at A 36 / B 53 ff.).

65. McTaggart, "The Unreality of Time"; see also McTaggart, *The Nature of Existence.*

66. For a position of this sort, see, e.g., Mellor, *Real Time II.*

67. In the course of a trenchant analysis, Cutrofello ("Revolutionary Actions and Events," 296) makes essentially this point on behalf of the paradoxico-critical position, also helpfully drawing out some further implications with respect to the critique of Badiou's generic orientation, as well as what McTaggart himself understood as a real third "C-series" underlying the (for him, illusory) A-series and B-series themselves.

68. The essentially formal character of this indication is sufficient to mark it off, both methodologically and thematically, from anything that could be produced or demonstrated by mean of an essential nonformal or unformalizable *intuition* of time or becoming in the manner of Bergson (in *Matter and Memory* [1896] and *Creative Evolution* [1907]). While there are some structural similarities between the current argument and Bergson's (particularly in relation to the structure of what he calls the "virtual"), what is most decisive in producing the current structure is the very possibility of *thinking* time as such and in general, which Bergson appears to deny.

69. For a relevant and insightful recent analysis of the difference between time and space, one of whose conclusions (210–11) is that subjectivity is dependent on time and space as limit-structures (rather than conversely), see McCumber, "Why Is Time Different from Space?"

70. This accordingly gives a realist answer to the problem about times prior to human subjectivity or language which is posed by Meillassoux in *After Finitude*, against (what he calls) "correlationist" positions and modes of thought.

71. Although the position that results from this is not identical with any position that has been (to my knowledge) formulated within recent "analytic" discussions of the metaphysics of time, it is probably most closely related to what has been called the "moving spotlight" account. On this account, realism about statements about the past and the future is maintained (and so "presentism" is denied), but it is also recognized that there is an "objectively privileged" present moment (and so it is possible to maintain realism about irreducibly temporal indexical judgments as well). For a convincing recent defense of the "moving spotlight" view, see Cameron, *The Moving Spotlight.* As Cameron argues, and as is confirmed from the current (ontologically motivated) perspective, it is better to think of the "moving spotlight" view as essentially combining presentism with the additional commitment of realism about past and future events, rather than as the combination of eternalism with an additional (and typically mysterious) privileging of *one* moment as the "objective" "now."

72. A further consideration in support of a realist construal of temporal indexicality, and ultimately in favor of a realist and paradoxico-critical attitude to it, is (ironically, given the title) brought out by Dummett himself in his 1960 article "A Defence of McTaggart's Proof of the Unreality of Time." Here Dummett argues that the defense of the reality of B-series temporal relations on the basis of con-

siderations about the behavior of indexicals fails, since it fails to account for what is actually an important *difference* between the functioning of temporal and spatial indexicals. In particular, it appears straightforwardly possible to imagine an observer describing the (spatial) position of all objects in space, although she does not (herself) occupy any particular position in the space described. By contrast with this, although we can picture an observer who is capable of giving a *complete* description of all temporal relations from a position that is itself not located anywhere in time, what such an observer would describe could only be a *static* representation of temporal relations and not those relations themselves. The description would thus inevitably falsify the phenomena and thus cannot be treated as complete with respect to them.

As Dummett concludes, this leaves us with just two alternatives: *either* to repudiate the reality of time in the sense of *both* the A-series *and* the B-series—this is McTaggart's own solution—*or*, while maintaining temporal realism, accordingly to repudiate the very idea of a position from which it is possible to give a *complete* description of temporally changing reality. Dummett opts cautiously for the first alternative. But as we have seen, there is a third alternative: that of maintaining temporal realism, including realism about change and becoming, *along with* the assumption of completeness by means of a positive ontological description that is ultimately paradoxical and contradictory in the ways that the constitutive paradoxes developed here bring out.

73. It is for this reason that I have not generally sought to apply to the current problematic a "tense logic" in the style of Prior (*Time and Modality*; *Past, Present, and Future*). As Prior convincingly argues, the use of such a logic to formalize temporal statements serves to clarify the logic of many of these statements and also to illuminate the underlying structure of temporal change and becoming itself. It can also be used to dissolve or eliminate McTaggart's paradox, since the contradiction which McTaggart argues is involved in characterizing A-series relations does not even arise if the temporal determinations involved in it are all expressed (at a particular time) in a tense logic (see *Past, Present, and Future,* 1–12). However, from the current perspective, the application of such a logic, with its own irreducible operators signifying pastness, presentness, and futurity, amounts essentially once more to a strategy of parameterization with respect to the actually unitary ontological phenomenon of given time. That is, the expression of temporal determinations in a tense logic still leaves mysterious the ultimate basis for the sense of the distinct temporal operators themselves, which is rather to be clarified by an analysis along the lines of the present one. (See, however, note 22 to chap. 7, below, for one suggestive and relevant argument formulable in Prior's tense logic.)

74. The first is at least implicit in Aristotle's claims that time is infinite in that it "never gives out" and that "time is always at a beginning and at an end" (222b4). The second is explicit at 218a9–218a29.

75. See, e.g., Plato, *Parmenides* 142b–145a; *Timaeus* 52a–b; and *Sophist* 244a–245d.

76. See, e.g., Plato, *Cratylus* 437a–440a; *Philebus* 24c–d; *Theaetetus* 181c–183a; and *Parmenides* 151e–152e. Compare also what Deleuze (*The Logic of Sense,*

2–3) treats as the paradoxes of an "unlimited becoming" in Plato which threatens to undermine all determinate identities, and for more discussion see appendix 2.

77. For an analysis of this problematic as it figures in Plato, see appendix 2.

78. Cf. the account of "worlds," their structuring "transcendentals," and the possibility of their transformation by means of a "retroaction" of appearance onto in Badiou's 2006 *Logics of Worlds.*

79. For a more detailed critical consideration of this picture and its relationship to the paradoxico-critical orientation, see Livingston, *The Politics of Logic,* chaps. 8 and 9.

80. Compare Badiou's account in *Being and Event* (210), where he suggests that time itself may be seen as determined by the (subjective) form of the intervention, which traces out the situational implications of an event.

81. In these terms and with appropriate qualifications, Badiou's picture of the (ontological) event and its relation to the distinct local situations and worlds it determines is simply a more formally rigorous (and perhaps somewhat more subjective-decisionist) development of Heidegger's own being-historical picture.

82. It is in this light that we should also apparently see the middle and later Heidegger's repudiations, beginning in the *Beiträge,* of any and all figures of "transcendence" in the context of the (later) question of the truth of beyng. See, e.g., GA 65, 217–18; and for more discussion, see appendix 1.

83. See chap. 6, sec. 1, above.

Chapter 7

1. It is, I believe, a consequence of this drive toward totalization that it is not as *critically* helpful to characterize the "logic" of contemporary capitalism in terms of its necessary incompleteness as it is to characterize it in terms of its claim to totality and the structuring inconsistencies (or antagonisms) that thereby become evident in it; thus as an overall response to the problems of contemporary capitalism, paradoxico-criticism is to be favored over the generic orientation. For a fuller argument for this point, see Livingston, *The Politics of Logic,* chap. 10.

2. Cf. Jacques Derrida in "différance" (22): "Perhaps we must attempt to think this unheard-of thought, this silent tracing: that the history of Being, whose thought engages the Greco-Western *logos* such as it is produced via the ontological difference, is but an epoch of the *diapherein.*"

3. The phenomenon of "*Machenschaft*" and the ubiquity of its determining force is centrally discussed, for example, in the *Beiträge* (GA 65) and the Nietzsche lectures of summer 1939 (GA 6, vol. 1), especially the two lectures on "The Eternal Recurrence of the Same and the Will to Power," which were written for the course but never delivered.

4. GA 7, 9–10.

5. GA 15, 366 (transl. slightly modified). A partially similar formulation is given in "The Principle of Identity" (1957), GA 11, 48–49; as well as already in the *Beiträge*: "Machination as the essential occurrence [*Wesung*] of beingness [*Seiendheit*] gives a first hint [*gibt einen ersten Wink*] of the truth of Beyng itself" (GA 65, 127).

6. For a helpful recent analysis of the idea of the singularity, see Chalmers, "The Singularity."

7. GA 14, 71–72.

8. GA 65, 108–9.

9. Ibid., 129.

10. Ibid., 136.

11. Ibid., 121 (transl. slightly modified).

12. Ibid., 120.

13. Turing, "On Computable Numbers."

14. More precisely, what Turing demonstrated is that *for any given* formal system or "universal" Turing machine, there will be well-defined decision problems that *that* system cannot solve. Whether this points simply to the limitation of each such system with respect to problems that can indeed be solved by an infinitary power of human thought irreducible to finite procedures, or whether it witnesses the real structure of problems unsolvable by *any* means or powers whatsoever, is just what is at issue in the dilemma of the "disjunctive" conclusion that Gödel draws in the 1953 Gibbs lecture (see chap. 5 above).

15. This is as opposed both to rational numbers but also to real numbers such as π or $\sqrt{2}$, whose successive digits *can* be determined by successive applications of such a rule.

16. In particular, the recursive enumerability of theorems in systems of the type Gödel considered.

17. For a fuller version of essentially this point, see Livingston, *The Politics of Logic*, chap. 10, and Livingston, "How Do We Recognize Strong Critique?"

18. Thomson, *Heidegger on Ontotheology*, 8–19 (quoted in chap. 1 above).

19. Schürmann, *Heidegger on Being and Acting*, 10.

20. GA 6, vol. 1, 296.

21. The assumptions of finitude in the course of events, and infinite repeatability in thought, which figure in this structure might be thought of as more sophisticated and formally/ontologically grounded version of ontic premises to which Nietzsche sometimes appeals in his own cosmological argumentation for eternal recurrence, e.g., those of the conservation of energy or action (finitude) and the determinism of physical laws (infinite repeatability of determination). Relatedly, see Deleuze, *Difference and Repetition*, for a development of an idea of thinkable repetition outside the constraint of identity, and a consequent development of Nietzsche's doctrine as in fact involving an eternal return of the different, or of "difference in itself."

22. There may seem to be a slide here from the finite *determination* of the course of the world *in thought* to the finitude of this course itself. However, for the present argument it may suffice simply to point out that a finitely *determined* course of the world will also be one which it is possible to *conceive* of as completed (even if this completion would take a [countably] infinite amount of time) and thus as already having taken place given infinite time in the past direction up to the present. The decisive consideration here would be that given infinite time in the past direction, "everything that *can* happen must already *have* happened." That is, given infinite past time, it is apparently impossible to think of the total-

ity of events leading up to the present as simply finite without repetition, since any finite (or indeed finitely determined) course of events would have to have (at some time in this infinite past) *already* run its course. (Kant appeals to this consideration in his argument, in the Antinomies, against an infinite past of the world, and it appears to be at least implicit in Nietzsche's argumentation as well.)

Admittedly, though, it is not completely clear that this consideration is itself justified. Why could there not be a total chain of events which is *thinkable* in finite form while nevertheless never repeating, and some of whose possibilities have already occurred while others have not yet? Thus (and has often been pointed out against arguments for recurrence) if the total series of *past and future* events were thinkable in the way an irrational number (such as π) is on the basis of the rule for its expansion, then it would be both finitely thinkable and of infinite extent, without the truth of recurrence apparently following from this.

On the other hand, however, while this is clearly an open possibility for the determination of *future* events *given* an initial "starting point" in the present, it is not as clear that it is coherent to suppose this form of determination operating in the *past* direction, as operating effectively "through all past time" and thus determining the *present* "now." Wittgenstein is reported (Moore, *The Infinite,* 44) to have once asked his audience in a lecture to imagine finding a man who has been reading out the digits of π, backward, for all time: he has just said, "3! I'm finished." What appears to be problematic here (or perhaps incoherent) is the idea of the determination of an actually infinite series *actually* being completed "in the backward direction" while also conditioning, as a totality, the present "now." (The anecdote appears to have been reported to Jonathan Bennett by Elizabeth Anscombe; cf. Bennett, "The Age and Size of the World," 135; and see also Wittgenstein, *Philosophical Remarks,* 166, for a partially similar example.) Nevertheless, it is not clear *why,* if such determination is indeed coherent in the "future" direction, it should be thus incoherent in the "past" one, and so this argument (if it is one), although suggestive, does not appear completely decisive in establishing the actual and literal truth of eternal recurrence. For some closely related argumentation and considerations, see Becker, "Nietzsche's Beweise."

23. I do not maintain that the reasoning in this paragraph and the last one represents, even in the context of broadly paradoxico-critical assumptions, an argument sufficient to *demonstrate* the (literal) *truth* of the doctrine of the recurrence of all (ontic) events. As it is invoked here, eternal recurrence is, at best, one way of *figuring* in a coherent picture the implications of the paradoxico-critical orientation, and it is not necessarily (as discussed below) the only possible one. Nevertheless, it is worth considering briefly what more *would* be needed in order to make it into a rigorous argument from the considerations about the availability of the thought of the totality of the world, on which the paradoxico-critical orientation turns, to the literal and exclusive truth of the cosmological doctrine.

At least a partial answer to this is suggested by Prior (*Past, Present, and Future,* 79–85), who, employing the resources of tense logic, rigorously considers the implications of adding to the basic resources of such a logic a single empirical premise (a "world-sentence") formulating the *current* total state of the world, including the totality of the *current* truths about it (cf. Wittgenstein's "The world is all that is the case"). Assuming the availability of such a premise, Prior formulates an argument with the conclusion that *if* the (current) total world-state has occurred at least once before, it will occur infinitely (countably) many times more, and indeed already has infinitely many times in the past. The argument thus rigorously establishes the *literal* truth of eternal recurrence given the availability of a total description of the world, but only on the additional premise of at least one past occurrence of the (current) total world-state. And it is, of course, not obvious how this additional premise itself could be verified.

24. GA 6, vol. 1, 318.

25. Ibid., 290.

26. Ibid., 293.

27. Ibid., 294.

28. Ibid., 295–96.

29. Ibid., 297.

30. Klossowski, *Nietzsche and the Vicious Circle*, 57, 58.

31. Ibid., 58.

32. Ibid.

33. Cf. Heidegger's 1962 lecture "Time and Being" (GA 14, 24): "In the sending of the destiny of Being [*im Schicken des Geschickes von Sein*], in the extending of time, there becomes manifest a dedication, a delivering over into what is their own [*ein Übereignen*], namely of Being as presence [*von Sein als Anwesenheit*] and of time as the realm of the open [*und von Zeit als Bereich des Offenen in ihr Eigenes*]. What determines both, time and Being, in their own [*in ihr Eigenes*], that is, in their belonging together, we shall call: *Ereignis*. What this word names can be thought now only in the light of what becomes manifest in our looking ahead toward Being and toward time as destiny and as extending, to which time and Being belong [*was sich in der Vor-Sicht auf Sein und auf Zeit als Geschick und als Reichen bekundet, wohin Zeit und Sein gehören*]."

34. GA 6, vols. 1 and 2. The latter two lectures are published as "The Eternal Recurrence of the Same and the Will to Power."

35. GA 6, vol. 1, 648.

36. GA 6, vol. 2, 11.

37. Ibid., 21 (transl. modified).

38. Compare Badiou, *The Century*, chap. 12.

39. The quotation is from a letter to Strindberg of 8 December 1888. For citation and some discussion, see Badiou, "Who Is Nietzsche?," 3–5.

40. Cf. Wittgenstein, *PI*, secs. 126, 435.

41. Cf. Deleuze, *Logic of Sense*, 71.

42. See, e.g., Marx, *Capital*, 637–38.

Bibliography

Texts by Heidegger

For texts by Heidegger, page number citations in text are to the German GA edition cited. Quoted texts in English are from the translation listed, unless otherwise noted.

GA 1 — *Frühe Schriften*. Frankfurt: Klostermann, 1972.

GA 2 — *Sein und Zeit*. 19te Auflage. Tübingen: Max Niemeyer Verlag, 2006. Translated by John Macquarrie and Edward Robinson as *Being and Time*. San Francisco: HarperSanFrancisco, 1962.

GA 3 — *Kant und das Problem der Metaphysik*. 2te unveränderte Auflage. Frankfurt: Klostermann, 1951. Reprint, Frankfurt: Klostermann, 1991. Translated by Richard Taft as *Kant and the Problem of Metaphysics*, 5th ed., enlarged. Bloomington and Indianapolis: Indiana University Press, 1997. (Page references in text are to the 1951 German edition.)

GA 5 — *Holzwege*. 7te durchgesehene Auflage. Frankfurt: Klostermann, 1994. Translated by Julian Young and Kenneth Haynes as *Off the Beaten Track*. Cambridge: Cambridge University Press, 2002.

GA 6, vol. 1 — *Nietzsche 1*. Pfullingen: Verlag Günther Neske Pfullingen, 1961. Partially translated by David F. Krell as *Nietzsche*, vol. 1: *The Will to Power as Art* and *Nietzsche*: vol. 2, *The Eternal Recurrence of the Same*. New York: HarperCollins, 1991.

GA 6, vol. 2 — *Nietzsche 2*. Frankfurt: Klostermann, 1997. Partially translated by Joan Stambaugh and Frank A. Capuzzi as *Nietzsche*, vol. 3: *The Will to Power as Knowledge and as Metaphysics* and *Nietzsche*, vol. 4: *Nihilism*. New York: HarperCollins, 1991.

GA 7 — *Vorträge und Aufsätze*. Frankfurt: Klostermann, 2000.

GA 9 — *Wegmarken*. 3te durchgesehene Auflage. Frankfurt: Klostermann, 1996. Translated by William McNeil as *Pathmarks*. Cambridge: Cambridge University Press, 1998.

GA 10 *Der Satz vom Grund.* Frankfurt: Klostermann, 1997. Translated by Reginald Lilly as *The Principle of Reason.* Bloomington and Indianapolis: Indiana University Press, 1996.

GA 11 *Identität und Differenz.* Frankfurt: Klostermann, 2006. Partially translated by Joan Stambaugh as *Identity and Difference.* Chicago: University of Chicago Press, 1969.

GA 12 *Unterwegs zur Sprache.* Frankfurt: Klostermann, 1985. Translated by Peter D. Herz as *On the Way to Language.* New York: Harper & Row, 1982.

GA 14 *Zur Sache des Denkens.* Frankfurt: Klostermann, 2007. Translated by Joan Stambaugh as *On Time and Being.* Chicago: University of Chicago Press, 2002.

GA 15 *Seminare.* 2te Auflage. Frankfurt: Klostermann, 2005. Partially translated by Andrew Mitchell and François Raffoul as *Four Seminars.* Bloomington and Indianapolis: Indiana University Press, 2003.

GA 19 *Platon: Sophistes.* Frankfurt: Klostermann, 1992. Translated by Richard Rojcewicz and André Schuwer as *Plato's Sophist.* Bloomington and Indianapolis: Indiana University Press, 2003.

GA 20 *Prolegomena zur Geschichte des Zeitbegriffs.* Frankfurt: Klostermann, 1979. Translated by Theodore Kisiel as *History of the Concept of Time: Prolegomena.* Bloomington and Indianapolis: Indiana University Press, 1992.

GA 21 *Logik: Die Frage nach der Wahrheit.* 2te Auflage. Frankfurt: Klostermann, 1995. Translated by Thomas Sheehan as *Logic: The Question of Truth.* Bloomington and Indianapolis: Indiana University Press, 2010.

GA 24 *Die Grundprobleme der Phänomenologie.* Frankfurt: Klostermann, 1975. Translated by Albert Hofstadter as *The Basic Problems of Phenomenology*, rev. ed. Bloomington and Indianapolis: Indiana University Press, 1982.

GA 26 *Metaphysische Anfangsgründe der Logik im Ausgang von Leibniz.* Frankfurt: Klostermann, 1978. Translated by Michael Heim as *The Metaphysical Foundations of Logic.* Bloomington and Indianapolis: Indiana University Press, 1984.

GA 34 *Vom Wesen der Wahrheit: Zu Platons Höhlengleichnis und Theätet.* 2te Auflage. Frankfurt: Klostermann, 1997.

GA 36/37 *Sein und Wahrheit.* Frankfurt: Klostermann, 2001. Translated by Gregory Fried and Richard Polt as *Being and Truth.* Bloomington and Indianapolis: Indiana University Press, 2010.

GA 38 *Logik als die Frage nach dem Wesen der Sprache.* Frankfurt: Klostermann, 1998. Translated by Wanda Torres Gregory and Yvonne Unna as *Logic as the Question concerning the Essence of Language.* Albany: SUNY Press, 2009.

GA 40 *Einführung in die Metaphysik.* 4te Auflage. Frankfurt: Klostermann, 1983. Translated by Gregory Field and Richard Polt as *Introduction to Metaphysics.* New Haven, Conn.: Yale University Press, 2000.

GA 56/57 *Zur Bestimmung der Philosophie.* 2te Auslegung. Frankfurt: Klostermann, 1999. Translated by Ted Sadler as *Towards the Definition of Philosophy.* London: Continuum, 2008.

GA 60 *Phänomenologie des religiösen Lebens.* 2te Auflage. Frankfurt: Klostermann, 2011. Translated by Matthias Fritsch and Jennifer Anna Gosetti-Ferencei as *Phenomenology of Religious Life.* Bloomington and Indianapolis: Indiana University Press, 2004.

GA 61 *Phänomenologische Interpretationen zu Aristoteles: Einführung in die phänomenologische Forschung.* 2te Auslage. Frankfurt: Klostermann, 1994. Translated by Richard Rojcewicz as *Phenomenological Interpretations of Aristotle: Initiation into Phenomenological Research.* Bloomington and Indianapolis: Indiana University Press, 2001.

GA 63 *Ontologie: Hermeneutik der Faktizität.* Frankfurt: Klostermann, 1988. Translated by John van Buren as *Ontology—The Hermeneutics of Facticity.* Bloomington and Indianapolis: Indiana University Press, 1999.

GA 65 *Beiträge zur Philosophie (Vom Ereignis).* Frankfurt: Klostermann, 1989. Translated by Richard Rojcewicz and Daniela Vallega-Neu as *Contributions to Philosophy: Of the Event.* Bloomington and Indianapolis: Indiana University Press, 2012.

GA 79 *Bremer und Freiburger Vorträge.* 2te Auflage. Frankfurt: Klostermann, 2005. Translated by Andrew J. Mitchell as *Bremen and Freiburg Lectures.* Bloomington and Indianapolis: Indiana University Press, 2012.

GA 85 *Vom Wesen der Sprache: Die Metaphysik der Sprache und die Wesen des Wortes: Zu Herders Abhandlung "Über den Ursprung der Sprache."* Frankfurt: Klostermann, 1999. Translated by Wanda Torres Gregory and Yvonne Umma as *On the Essence of Language: The Metaphysics of Language and the Essencing of the Word; Concerning Herder's Treatise on the Origin of Language.* Albany: SUNY Press, 2004.

Other Texts

Aristotle. *Complete Works of Aristotle: The Revised Oxford Translation.* Edited by J. Barnes. 2 vols. Bollingen Series. Princeton, N.J.: Princeton University Press, 1983.

Badiou, Alain. *Being and Event.* Translated by Oliver Feltham. London: Continuum, 2005.

———. *Briefings on Existence: A Short Treatise on Transitory Ontology.* Translated by Norman Madarasz. Albany: SUNY Press, 2006.

———. *The Century.* Translated by Alberto Toscano. Cambridge: Polity, 2007.

———. *Logics of Worlds: Being and Event II.* Translated by Alberto Toscano. London: Continuum, 2009.

———. "Who Is Nietzsche?" *PIi* 11 (2001): 1–11.

Beaney, Michael. "Introduction." In *The Frege Reader.* Oxford: Blackwell, 1997.

Becker, Oskar. *Mathematische Existenz.* Tübingen: Max Niemeyer Verlag, 1973.

———. "Nietzsche's Beweise für seine Lehre von der ewigen Wiederkehr." In *Dasein und Dawesen: Gesammelte Philosophische Aufsätze.* Pfullingen: Neske, 1963.

Bennett, Jonathan. "The Age and Size of the World." *Synthese* 23, no. 1 (1971): 127–46.

Bergson, Henri. *Creative Evolution.* Translated by Arthur Mitchell. New York: Dover, 1998.

———. *Matter and Memory.* Translated by N. M. Paul and W. S. Palmer. New York: Zone Books, 1994.

Berto, Francesco. *There's Something about Gödel: The Complete Guide to the Incompleteness Theorem.* Hoboken, N.J.: Wiley-Blackwell, 2009.

Blattner, William. *Heidegger's Temporal Idealism.* Cambridge: Cambridge University Press, 1999.

Bova, John. "A Metalogical Approach to Platonic Dialectic." Ph.D. diss., Villanova University, 2016.

———. "Negation and Incommensurability: On the Metalogical Kernal of the Platonic Dialectic." Talk at Villanova University, 28 March 2010. Abstract available at https://www.academia.edu/224135/Negation_Incommensurability_Abstract_.

Brandom, Robert. *Articulating Reasons: An Introduction to Inferentialism.* Cambridge, Mass.: Harvard University Press, 2001.

———. *Making It Explicit: Reasoning, Representing, and Discursive Commitment.* Cambridge, Mass.: Harvard University Press, 1994.

Braver, Lee. *Groundless Grounds: A Study of Wittgenstein and Heidegger.* Cambridge, Mass.: MIT Press, 2012.

Brogan, Walter. *Heidegger and Aristotle: The Twofoldness of Being.* Albany: SUNY Press, 2006.

Cameron, Ross P. *The Moving Spotlight: An Essay on Time and Ontology.* Oxford: Oxford University Press, 2015.

Carman, Taylor. *Heidegger's Analytic: Interpretation, Discourse, and Authenticity in* Being and Time. Cambridge: Cambridge University Press, 2003.

Carnap, Rudolf. *Meaning and Necessity: A Study in Semantics and Modal Logic.* Chicago: University of Chicago Press, 1947.

Cavell, Stanley. *The Claim of Reason: Wittgenstein, Skepticism, Morality, and Tragedy.* New York: Oxford University Press, 1999.

———. *Must We Mean What We Say? A Book of Essays.* Updated ed. New York: Cambridge University Press, 2002.

Chalmers, David J. *The Conscious Mind: In Search of a Fundamental Theory.* Oxford: Oxford University Press, 1996.

———. "The Singularity: A Philosophical Analysis." *Journal of Consciousness Studies* 17 (2010): 7–65.

Church, Alonzo. *Introduction to Mathematical Logic.* Vol. 1. Princeton, N.J.: Princeton University Press, 1956.

Crowell, Steven Galt. *Husserl, Heidegger, and the Space of Meaning.* Evanston, Ill.: Northwestern University Press, 2001.

Cutrofello, Andrew. "Revolutionary Actions and Events." In *Beyond the Analytic-Continental Divide: Pluralist Philosophy in the Twenty-First Century,* edited by J. A. Bell, A. Cutrofello, and P. M. Livingston, 287–304. New York: Routledge, 2016.

Dahlstrom, Daniel O. *Heidegger's Concept of Truth.* New York: Cambridge University Press, 2001.

———. "Heidegger's Method: Philosophical Concepts as Formal Indications." *Review of Metaphysics* 47, no. 4 (1994): 775–95.

Davidson, Donald. "Belief and the Basis of Meaning." In *Inquiries into Truth and Interpretation.*

———. "The Folly of Trying to Define Truth." *Journal of Philosophy* 93, no. 6 (1996): 263–78.

———. *Inquiries into Truth and Interpretation.* 2nd ed. Oxford: Clarendon, 2001.

———. "On the Very Idea of a Conceptual Scheme." In *Inquiries into Truth and Interpretation.*

———. "Radical Interpretation." In *Inquiries into Truth and Interpretation.*

———. "Semantics for Natural Languages." In *Inquiries into Truth and Interpretation.*

———. "Theories of Meaning and Learnable Languages." In *Inquiries into Truth and Interpretation.*

———. "Truth and Meaning." In *Inquiries into Truth and Interpretation.*

———. *Truth and Predication.* Cambridge, Mass.: Harvard University Press, 2005.

Deleuze, Gilles. *Difference and Repetition.* Translated by Paul Patton. New York: Columbia University Press, 1994.

———. *The Logic of Sense.* Translated by Mark Lester. New York: Columbia University Press, 1990.

Derrida, Jacques. "différance." In *Margins of Philosophy.* Translated by Alan Bass. Chicago: University of Chicago Press, 1982.

———. "*Ousia* and *Grammē*: Note on a Note from *Being and Time.*" In *Margins of Philosophy.* Translated by Alan Bass. Chicago: University of Chicago Press, 1982.

Dreyfus, Hubert. *Being-in-the-World: A Commentary on Heidegger's* Being and Time, *Division 1.* Cambridge, Mass.: MIT Press, 1990.

———. "Comments on Cristina Lafont's Interpretation of *Being and Time.*" *Inquiry* 45, no. 2 (2002): 191–94.

Dummett, Michael. "Can Truth Be Defined?" In *Frege: Philosophy of Language.* 2nd ed. London: Duckworth, 1981.

———. "A Defence of McTaggart's Proof of the Unreality of Time." In *Truth and Other Enigmas.* London: Duckworth, 1978.

———. *Frege and Other Philosophers.* Oxford: Clarendon, 1991.

———. *Frege: Philosophy of Language.* 2nd ed. Cambridge, Mass.: Harvard University Press, 1981.

———. *The Interpretation of Frege's Philosophy.* London: Duckworth, 1981.

———. "The Philosophical Basis of Intuitionistic Logic." In *Truth and Other Enigmas.* London: Duckworth, 1978.

———. "The Philosophical Significance of Gödel's Theorem." In *Truth and Other Enigmas.* London: Duckworth, 1978.

———. "Preface." In *Truth and Other Enigmas.* London: Duckworth, 1978.

———. "Realism." In *Truth and Other Enigmas.* London: Duckworth, 1978.

———. "The Reality of the Past." In *Truth and Other Enigmas.* London: Duckworth, 1978.

———. "Truth." In *Truth and Other Enigmas.* London: Duckworth, 1978.

———. "What Is a Theory of Meaning? (I)." In *The Seas of Language.* Oxford: Clarendon, 1993.

Eyers, Tom. *Lacan and the Concept of the "Real."* New York: Palgrave Macmillan, 2012.

Feferman, Solomon. "Are There Absolutely Unsolvable Problems? Gödel's Dichotomy." *Philosophia Mathematica* 14, no. 2 (2006): 134–52 (page references in text are to online version at http://math.stanford.edu/~feferman/papers/dichotomy.pdf).

———"Transfinite Recursive Progressions of Axiomatic Theories." *Journal of Symbolic Logic* 27 (1962): 259–316.

Føllesdal, Dagfinn. "Husserl's Notion of Noema." In *Husserl, Intentionality, and Cognitive Science,* edited by Hubert Dreyfus. Cambridge, Mass.: MIT Press, 1982.

Frege, Gottlob. *Begriffsschrift, eine der arithmetischen nachgebildete Formelsprache des reinen Denkens.* Halle: L. Nebert, 1879. Selections translated by Michael Beaney in *The Frege Reader,* edited by Michael Beaney. Oxford: Blackwell, 1997.

———. "A Brief Survey of My Logical Doctrines: Extract." Translated by Peter Long and Roger White. In *The Frege Reader,* edited by Michael Beaney. Oxford: Blackwell, 1997.

———. "Function and Concept." Translated by Peter Geach. In *The Frege Reader,* edited by Michael Beaney. Oxford: Blackwell, 1997.

———. *Grundgesetze der Arithmetik.* Jena: H. Pohle, 1903. Translated by Philip A. Ebert and Marus Rossberg with Crispin Wright as *Basic Laws of Arithmetic: Derived Using Concept-Script.* Oxford: Oxford University Press, 2013.

———. "Logic." Translated by Peter Long and Roger White. In *The Frege Reader,* edited by Michael Beaney. Oxford: Blackwell, 1997.

———. "On *Sinn* and *Bedeutung.*" Translated by Max Black. In *The Frege Reader,* edited by Michael Beaney. Oxford: Blackwell, 1997.

———. "Thought." Translated by Peter Geach and R. H. Stoothoff. In *The Frege Reader,* edited by Michael Beaney. Oxford: Blackwell, 1997.

Gödel, Kurt. "Russell's Mathematical Logic." In *The Philosophy of Bertrand Russell*, edited by P. A. Schlipp. Evanston, Ill.: Northwestern University Press, 1944.

———. "Some Basic Theorems on the Foundations of Mathematics and Their Philosophical Implications." In *Kurt Gödel: Unpublished Philosophical Essays*, edited by F. A. Rodriguez-Consiguera. Basel: Springer, 1995.

Grim, Patrick. "There Is No Set of All Truths." *Analysis* 44 (1984): 206–8.

———. "What Is a Contradiction?" In *The Law of Non-Contradiction*, edited by Graham Priest, J. C. Beall, and Bradley Armour-Garb. Oxford: Clarendon, 2007.

Gupta, Anil, and Nuel Belnap. *The Revision Theory of Truth*. Cambridge, Mass.: MIT Press, 1993.

Hägglund, Martin. *Radical Atheism: Derrida and the Time of Life*. Stanford, Calif.: Stanford University Press, 2008.

Hanzel, Igor. "Frege, the Identity of Sinn, and Carnap's Intension." *History and Philosophy of Logic* 27, no. 3 (2006): 229–47.

Harman, Graham. *Tool-Being: Heidegger and the Metaphysics of Objects*. Chicago: Open Court, 2002.

Heck, Richard. "Predicative Fragments of the Formal Theory of Frege's *Grundgesetze der Arithmetik*." *History and Philosophy of Logic* 17 (1996): 209–20.

Husserl, Edmund. *Cartesian Meditations: An Introduction to Phenomenology*. Translated by Dorion Cairns. Dordrecht: Kluwer, 1999.

———. *Experience and Judgement*. Translated by J. S. Churchill and K. Ameriks. London: Routledge, 1973.

———. *Formal and Transcendental Logic*. Translated by Dorian Cairns. The Hague: Nijhoff, 1969.

———. *Ideas Pertaining to a Pure Phenomenology and to a Phenomenological Philosophy—First Book: General Introduction to a Pure Phenomenology*. Translated by F. Kersten. The Hague: Nijhoff, 1982.

———. *Logische Untersuchungen*. Halle: 1901; rev. ed. 1922. The Hague, Netherlands: Martinus Nijhoff, 1984. Translated by J. N. Findlay as *Logical Investigations*. 2 vols. London: Routledge and Kegan Paul, 1970.

Incurvati, Luca, and Julien Murzi. "Maximally Consistent Sets of Instances of Naïve Comprehension." *Mind* (forthcoming).

Kant, Immanuel. *Kritik der reinen Vernunft*. Hamburg: Felix Meiner, 1954. Translated by Paul Guyer and Allen W. Wood as *Critique of Pure Reason*. Cambridge: Cambridge University Press, 1998.

Kirk, Geoffrey S., John E. Raven, and Malcolm Schofield. *The Pre-Socratic Philosophers*. 2nd ed. Cambridge: Cambridge University Press, 1983.

Kisiel, Theodore. *The Genesis of Heidegger's Being and Time*. Berkeley and Los Angeles: University of California Press, 1993.

———. "Why Students of Heidegger Will Have to Read Emil Lask." *Man and World* 28, no. 3 (1995): 197–240.

Klossowski, Pierre. *Nietzsche and the Vicious Circle*. Translated by Daniel W. Smith. London: Athlone, 1997.

Kripke, Saul. *Naming and Necessity*. Cambridge, Mass.: Harvard University Press, 1980.

———. "Outline of a Theory of Truth." *Journal of Philosophy* 72 (1975): 690–716.

———. *Wittgenstein on Rules and Private Language.* Cambridge, Mass.: Harvard University Press, 1982.

Kuhn, Thomas S. "Reflections on My Critics." In *Criticism and the Growth of Knowledge,* edited by I. Lakatos and A. Musgrave. Cambridge: Cambridge University Press, 1970.

———. *The Structure of Scientific Revolutions.* 3rd ed. Chicago: University of Chicago Press, 1996.

Lacan, Jacques. *Encore: The Seminar of Jacques Lacan Book XX.* Edited by Jacques-Alain Miller. Translated by Bruce Fink. New York: Norton & Co., 1999.

Lafont, Cristina. *Heidegger, Language, and World-Disclosure.* Translated by Graham Harman. Cambridge: Cambridge University Press, 2000.

———. "Replies." *Inquiry* 45, no. 2 (2002): 229–48.

Lautman, Albert. "Essay on the Notions of Structure and Existence in Mathematics." Translated by Simon B. Duffy. In *Mathematics, Ideas and the Physical Real,* 87–193. London: Continuum, 2011.

———. "New Research on the Dialectical Structure of Mathematics." Translated by Simon B. Duffy. In *Mathematics, Ideas and the Physical Real.* London: Continuum, 2011.

Lewis, David K. "An Argument for the Identity Theory." *Journal of Philosophy* 63 (1966): 17–25.

———. "Psychophysical and Theoretical Identifications." *Australasian Journal of Philosophy,* 50 (1972): 249–58.

Livingston, Paul M. "Derrida and Formal Logic: Formalizing the Undecidable." *Derrida Today* 3, no. 2 (2010): 221–39.

———. "How Do We Recognize Strong Critique?" *Crisis and Critique* 3 (2014): 85–115.

———. "Lacan, Deleuze, and the Consequences of Formalism." In *Deleuze and Lacan: A Disjunctive Synthesis,* edited by Boštjan Nedoh and Andreja Zevnik. Edinburgh: Edinburgh University Press, forthcoming.

———. "Metaformal Realism and the Ontological Problematic." Talk at Fifth Annual Southwest Seminar in Continental Philosophy, Colorado College, Colorado Springs, Colo., June 2014.

———. *Philosophical History and the Problem of Consciousness.* New York: Cambridge University Press, 2004.

———. *Philosophy and the Vision of Language.* New York: Routledge, 2008.

———. *The Politics of Logic: Badiou, Wittgenstein, and the Consequences of Formalism.* New York: Routledge, 2012.

———. "The Sense of Finitude and the Finitude of Sense." In *Semantics and Beyond: Philosophical and Linguistic Investigations,* edited by Piotr Stalmaszczyk. Berlin: Ontos Verlag, 2014.

———. "Thinking and Being: Heidegger and Wittgenstein on Machination and Lived Experience." *Inquiry* 46, no. 3 (2003): 324–45.

Longuenesse, Béatrice. *Kant and the Capacity to Judge: Sensibility and Discursivity in the Transcendental Analytic of the Critique of Pure Reason.* Princeton, N.J.: Princeton University Press, 1998.

Malpas, Jeff. *Heidegger's Topology: Being, Place, World.* Cambridge, Mass.: MIT Press, 2006.

Martin, Wayne. *Theories of Judgment: Psychology, Logic, Phenomenology*. Cambridge: Cambridge University Press, 2006.

Marx, Karl. *Capital: A Critique of Political Economy*. Vol. 1. Translated by Ben Fowkes. London: Penguin, 1976.

McCumber, John. "Why Is Time Different from Space?" In *Beyond the Analytic-Continental Divide: Pluralist Philosophy in the Twenty-First Century*, edited by Jeffrey A. Bell, Andrew Cutrofello, and Paul M. Livingston. New York: Routledge, 2016.

McDaniel, Kris. "Ways of Being." In *Metametaphysics: New Essays on the Foundations of Ontology*, edited by David J. Chalmers, David Manley, and Ryan Wasserman. Oxford: Clarendon, 2009.

McDowell, John. *Mind and World*. Cambridge, Mass.: Harvard University Press, 1994.

McGee, Van. "Maximal Consistent Sets of Instances of Tarski's Schema (T)." *Journal of Philosophical Logic* 21 (1992): 235–41.

McTaggart, John M. E. *The Nature of Existence*. Vol. 2. New York: Cambridge University Press, 1988.

———. "The Unreality of Time." *Mind* 17, no. 68 (1908): 457–74.

Meillassoux, Quentin. *After Finitude: An Essay on the Necessity of Contingency*. Translated by Ray Brassier. London: Continuum, 2008.

Melamed, Yitzhak. "'Omnis determinatio est negatio'—Determination, Negation and Self-Negation in Spinoza, Kant, and Hegel." In *Spinoza and German Idealism*, edited by Eckart Förster and Yitzhak Y. Melamed. Cambridge: Cambridge University Press, 2012.

Mellor, D. H. *Real Time II*. London: Routledge, 1998.

Moore, Adrian W. *The Infinite*. 2nd ed. London: Routledge, 1990.

Okrent, Mark. *Heidegger's Pragmatism: Understanding, Being, and the Critique of Metaphysics*. Ithaca, N.Y.: Cornell University Press, 1988.

Perry, John. "The Problem of the Essential Indexical." *Noûs* 13 (1979): 3–21.

Plato. *Plato: Complete Works*. Edited by John M. Cooper. Indianapolis: Hackett, 1997.

———. "Sophist." Translated by Nicholas P. White. In *Plato: Complete Works*. Indianapolis: Hackett, 1997.

———. "Sophist." Translated by Harold North Fowler. In *Plato: Theaetetus; Sophist*. Loeb Classical Library. Cambridge, Mass.: Harvard University Press, 1996.

Priest, Graham. "The Answer to the Question of Being." In *Beyond the Analytic-Continental Divide: Pluralist Philosophy in the Twenty-First Century*, edited by Jeffrey A. Bell, Andrew Cutrofello, and Paul M. Livingston. New York: Routledge, 2016.

———. *Beyond the Limits of Thought*. 2nd ed. Oxford: Clarendon, 2002.

———. *Doubt Truth to Be a Liar*. Oxford: Clarendon, 2006.

———. *In Contradiction: A Study of the Transconsistent*. Exp. ed. Oxford: Oxford University Press, 2006.

———. *One: Being an Investigation into the Unity of Reality and of Its Parts, including the Singular Object Which Is Nothingness*. Oxford: Clarendon, 2014.

———. *Towards Non-Being: The Logic and Metaphysics of Intentionality*. Oxford: Clarendon, 2005.

Prior, A. N. *Past, Present, and Future*. Oxford: Clarendon, 1967.

———. *Time and Modality.* Oxford: Clarendon, 1957.

Quine, W. V. "Truth by Convention." In *The Ways of Paradox and Other Essays.* New York: Random House, 1966.

———. "Two Dogmas of Empiricism." *Philosophical Review* 60 (1951): 20–43.

Ramsey, Frank P. "Facts and Propositions." *Proceedings of the Aristotelian Society* 7, no. 1 (1927): 153–70.

Rorty, Richard. "Heidegger, Contingency, and Pragmatism." In *Essays on Heidegger and Others: Philosophical Papers*, vol. 2. Cambridge: Cambridge University Press, 1991.

Routley, Richard. *Exploring Meinong's Jungle and Beyond.* Canberra: RSSS, Australian National University, 1980.

Russell, Bertrand. *A Critical Exposition of the Philosophy of Leibniz.* London: Routledge, 1992.

———. "On Denoting." *Mind* 14 (1905): 479–93.

Schürmann, Reiner. *Broken Hegemonies.* Translated by Reginald Lilly. Bloomington: Indiana University Press, 2003.

———. *Heidegger on Being and Acting: From Principles to Anarchy.* Translated by Christine-Marie Gros. Bloomington: Indiana University Press, 1987.

Sellars, Wilfrid. "Empiricism and the Philosophy of Mind." In *Minnesota Studies in the Philosophy of Science* 1 (1956): 253–329.

Shapiro, Stewart. "Incompleteness, Mechanism, and Optimism." *Bulletin of Symbolic Logic* 4, no. 3 (1998): 273–302.

Shockey, R. Mathew. "What's Formal about Formal Indication? Heidegger's Method in *Sein und Zeit.*" *Inquiry* 53, no. 6 (2011): 525–39.

Smart, J. J. C. "Sensations and Brain Processes." *Philosophical Review* 68 (1959): 141–56.

Smith, David W. "Truth and Epoché: The Semantic Conception of Truth in Phenomenology." In *Beyond the Analytic-Continental Divide: Pluralist Philosophy in the Twenty-First Century*, edited by Jeffrey A. Bell, Andrew Cutrofello, and Paul M. Livingston. New York: Routledge, 2016.

Smith, William H. "Why Tugendhat's Critique of Heidegger's Concept of Truth Remains a Critical Problem." *Inquiry* 50, no. 2 (2007): 156–79.

Tarski, Alfred. "The Concept of Truth in Formalized Languages." In *Logic, Semantics, Metamathematics: Papers from 1923 to 1938.* 2nd ed. Translated by J. H. Woodger. Hackett, 1983.

———. "The Semantic Conception of Truth." *Philosophy and Phenomenological Research* 4, no. 3 (1944): 341–75.

Thomson, Iain. *Heidegger on Ontotheology: Technology and the Politics of Education.* Cambridge: Cambridge University Press, 2005.

Tugendhat, Ernst. *Der Wahrheitsbegriff bei Husserl und Heidegger.* Berlin: Walter de Gruyter & Co., 1967.

———. "Heidegger's Idea of Truth." In *The Heidegger Controversy: A Critical Reader.* Edited by Richard Wolin. New York: Columbia University Press, 1991.

Turing, Alan. "On Computable Numbers, with an Application to the Entscheidungsproblem." In *The Essential Turing*, edited by B. J. Copeland. Oxford: Clarendon, 2004.

van Buren, John. "The Ethics of Formale Anzeige in Heidegger." *American Catholic Philosophical Quarterly* 69, no. 2 (1995): 157–70.

———. *The Young Heidegger: Rumor of the Hidden King.* Bloomington: Indiana University Press, 1994.

Wittgenstein, Ludwig. *Philosophical Investigations.* Rev. 4th ed. Translated by G. E. M. Anscombe, P. M. S. Hacker, and Joachim Schulte. Oxford: Blackwell, 2009.

———. *Philosophical Remarks.* Edited by Rush Rhees and translated by Raymond Hargreaves and Roger White. Chicago: University of Chicago Press, 1975.

———. *Tractatus Logico-Philosophicus.* Translated by D. F. Pears and B. F. McGuinness. London: Routledge, 2001.

Wrathall, Mark. "Heidegger and Truth as Correspondence." *International Journal of Philosophical Studies* 7, no. 1 (1999): 69–88.

———. *Heidegger and Unconcealment: Truth, Language, and History.* Cambridge: Cambridge University Press, 2011.

Index